AF885510

SOTHIS PRESS

Sothis Press aims to completely reimagine the public domain book with an emphasis on universal access to captivatingly designed print editions. We believe that technology will enable every single book ever published to be perpetually accessible in beautiful print form. Sothis is beauty in ubiquity.

www.sothis.press
1 Waterhouse Square, London, ECIN 2ST

# The Geology of the Country Around Henley-on-Thames and Wallingford

A. J. Jukes-Browne

SOTHIS PRESS

ISBN 978-1-0365-1840-0

MEMOIRS OF THE GEOLOGICAL SURVEY.

ENGLAND AND WALES.

**EXPLANATION OF SHEET 254.**

# THE GEOLOGY OF THE COUNTRY AROUND HENLEY-ON-THAMES AND WALLINGFORD.

BY

A. J. JUKES-BROWNE, B.A., F.G.S.,

AND

H. J OSBORNE WHITE, F.G.S.

PUBLISHED BY ORDER OF THE LORDS COMMISSIONERS OF HIS MAJESTY'S TREASURY.

LONDON:

PRINTED FOR HIS MAJESTY'S STATIONERY OFFICE

BY WYMAN & SONS, LIMITED, FETTER LANE, E.C.

And to be purchased from

E. STANFORD, 12, 13, and 14, LONG ACRE, LONDON;

W. & A. K. JOHNSTON, LTD., 2, ST. ANDREW SQUARE, EDINBURGH;

HODGES, FIGGIS & Co., GRAFTON STREET, DUBLIN;

From any Agent for the sale of Ordnance Survey Maps; or through any Bookseller from the Ordnance Survey Office, Southampton.

1908.

*Price Two Shillings.*

MEMOIRS OF THE GEOLOGICAL SURVEY.

ENGLAND AND WALES.

**EXPLANATION OF SHEET 254.**

# THE GEOLOGY OF THE COUNTRY AROUND HENLEY-ON-THAMES AND WALLINGFORD.

BY

A. J. JUKES-BROWNE, B.A., F.G.S.,

AND

H. J OSBORNE WHITE, F.G.S.

PUBLISHED BY ORDER OF THE LORDS COMMISSIONERS OF HIS MAJESTY'S TREASURY.

LONDON:

PRINTED FOR HIS MAJESTY'S STATIONERY OFFICE
BY WYMAN & SONS, LIMITED, FETTER LANE, E.C.

And to be purchased from
E. STANFORD, 12, 13, and 14, LONG ACRE, LONDON;
W. & A. K. JOHNSTON, LTD., 2, ST. ANDREW SQUARE, EDINBURGH;
HODGES, FIGGIS & Co., GRAFTON STREET, DUBLIN;
From any Agent for the sale of Ordnance Survey Maps; or through any
Bookseller from the Ordnance Survey Office, Southampton.

1908.

*Price Two Shillings.*

# PREFACE.

The original geological survey of the area was made on the old series one-inch maps, and published on Sheets 7 and 13 (without drift) in 1861 and 1859, by E. Hull, H. Bauerman and W. Whitaker, and on Sheet 7 (with drift) in 1871, by H. B. Woodward, J. H. Blake and F. J. Bennett.

The area was re-surveyed on the six-inch scale by the late J. H. Blake and Mr. A. J. Jukes-Browne, and published on the new series one-inch map, Sheet 254, in 1905.

The geology of the area has been described in the separate memoirs on Sheets 7 and 13, in the larger memoirs on the Geology of the London Basin and Geology of London by Mr. Whitaker, and in the Memoir on the Cretaceous Rocks of Britain, by Mr. Jukes-Browne. The north-western portion of the area has been more recently described in the explanatory memoir accompanying the Special Oxford Sheet, and some modifications in the grouping of the strata (as noted on p. 7), have been made on that map.

The area includes a considerable portion of the Chiltern Hills, and part of the Thames Valley.

The superficial deposits and their relation to the scenery have perhaps attracted most attention from geologists, from the days of Buckland and Conybeare, to those of Prestwich, who has been followed by Messrs. R. S. Herries, H. W. Monckton A. E. Salter, O. A. Shrubsole, Osborne White, and others.

The Cretaceous chapters of the present memoir were written by Mr. Jukes-Browne, prior to his retirement from the Geological Survey; and his observations on the zonal divisions in the Chalk have been supplemented by those of Mr. Osborne White and Mr. Llewelyn Treacher.

Short notes on the Jurassic and Lower Cretaceous formations are contributed by Mr. H. B. Woodward, and Mr. G. W. Lamplugh.

For the chapters on the Eocene and superficial deposits we are indebted to Mr. Osborne White, who has incorporated notes made by the late J. H. Blake.

Mr. H. H. Thomas has contributed some notes on rock-specimens, and we are indebted to Mr. A. S. Kennard for identifying specimens of mollusca from the Alluvial deposits.

J. J. H. TEALL,
Director.

*Geological Survey Office,*
*28, Jermyn Street,*
*London.*

5 August, 1908.

10322. 500.—Wt. 29091/ 9/08.—Wy. & S

## NOTE.

The following is a list of the six-inch Geological Maps included in the area, of which MS. coloured copies are deposited for public reference in the Library of the Geological Survey and Museum of Practical Geology :—

### BERKSHIRE.

| | | |
|---|---|---|
| 15 N.E. | Didcot. | By A. J. Jukes-Browne, 1886 |
| 15 S.E. | East Hagbourne. | ,, ,, 1886 |
| 16 S.W. | N. & S. Moreton. | ,, ,, 1886 |
| 21 N.E. | Blewbury. | ,, ,, 1887 |
| 21 S.E. | Aston Upthorpe Downs. | By A. J. Jukes-Browne and F. J. Bennett, 1885-87 |
| 23 | (Bucks. 51, Oxfordshire 54) Henley-on-Thames | By J. H. Blake, 1896 |

### OXFORDSHIRE.

| | | |
|---|---|---|
| 39 S.E. | (Berks. 6 S.E.) N. of Radley. | By J. H. Blake, 1899, and T. I. Pocock, 1905 |
| 40 S.W. | Toot Baldon. | By J. H. Blake, 1899, and H. B. Woodward, 1907 |
| 40 S.E. | Hangman's Bridge. | By J. H. Blake |
| 41 S.W. | N. Wheatfield Park. | ,, |
| 41 S.E. | Postcombe. | By A. J. Jukes-Browne, 1885-86. |
| 42 S.W. | (Bucks. 37 S.W.) Chinnor Hill. | By A. J. Jukes-Browne, 1887, and J. H. Blake |
| 45 N.E. | (Berks. 10 N.E.) Nuneham Courtenay. | By A. J. Jukes-Browne, J. H. Blake, 1897, T. I. Pocock, 1905, and H. B. Woodward, 1907 |
| 45 S.E. | (Berks. 10 S.E.) Clifton Hampden | By A. J. Jukes-Browne, 1887, and J. H. Blake, 1897 |
| 46 N.W. | Chislehampton | By A. J. Jukes-Browne, 1887, J. H. Blake, 1899, and H. B. Woodward, 1907 |
| 46 N.E. | Chalgrove, Ascot. | By J. H. Blake, 1899 |
| 46 S.W. | (Berks. 11 S.W.) Dorchester. | By A. J. Jukes-Browne, 1886, and J. H. Blake, 1899 |
| 46 S.E. | Berrick Salome. | By A. J. Jukes-Browne, 1886 |
| 47 N.W. | Stoke Talmage. | By A. J. Jukes-Browne, 1886, and J. H. Blake, 1899 |
| 47 S.W. | Watlington. | By A. J. Jukes-Browne, 1886 |
| 49 N.W. | (Berks. 16 N.W.) Brightwell. | ,, ,, 1886 |
| 49 N.E. | ( ,, 16 N.E.) Ewelme. | ,, ,, 1886 |
| 49 S.E. | ( ,, 16 S.E.) Wallingford. | ,, ,, 1886 |
| 50 N.W. | Britwell Salome. | ,, ,, 1886 |

| | | |
|---|---|---|
| 50 S.W. | Nuffield. | By A. J. Jukes-Browne, 1886, and J. H. Blake |
| 52 N.W. | (Berks. 22 N.W.) Moulsford | By A. J. Jukes-Browne, 1886. |
| 52 N.E. | ( ,, 22 N.E.) Ipsden. | ,, ,, 1886. |
| 52 S.W. | ( ,, 22 S.W.) Goring. | By A. J. Jukes-Browne and J. H. Blake, 1891 |
| 52 S.E. | Woodcote. | By J. H. Blake, 1889 |
| 53 | (Bucks. 50) Stoke Row, Nettlebed. | ,, 1896 |

BUCKINGHAMSHIRE.

| | | |
|---|---|---|
| 37 S.E. | N. of Saunderton Lee. | By A. J. Jukes-Browne, 1885 and J. H. Blake |
| 40 N.E. | (Oxford 47 N.E.) Aston Rowant. | By A. J. Jukes-Browne, 1886, and J. H. Blake |
| 40 S.E. | (Oxford 47 S.E.) Shirburn Hill. | By A. J. Jukes-Browne, 1886, and J. H. Blake, 1899 |
| 41 N.W. | (Oxford 48 N.W.) Radnage. | ,, ,, 1899 |
| 41 N.E. | Saunderton Lee. | By J. H. Blake |
| 41 S.W. | Stokenchurch. | ,, |
| 41 S.E. | Chawley Manor Farm. | ,, |
| 45 N.E. | (Oxford 50 N.E.) Northend. | ,, 1891 |
| 45 S.E. | ( ,, 50 S.E.) Stonor Park. | ,, 1890 |
| 46 N.W. | Fingest, Cadmore End. | ,, 1900 |
| 46 N.E. | Lane End. | ,, 1900 |
| 46 S.W. | (Oxford 51 S.W.) Skirmett. | ,, 1900 |
| 46 S.E. | Moor End. | ,, 1900 |

# CONTENTS.

# ILLUSTRATIONS.

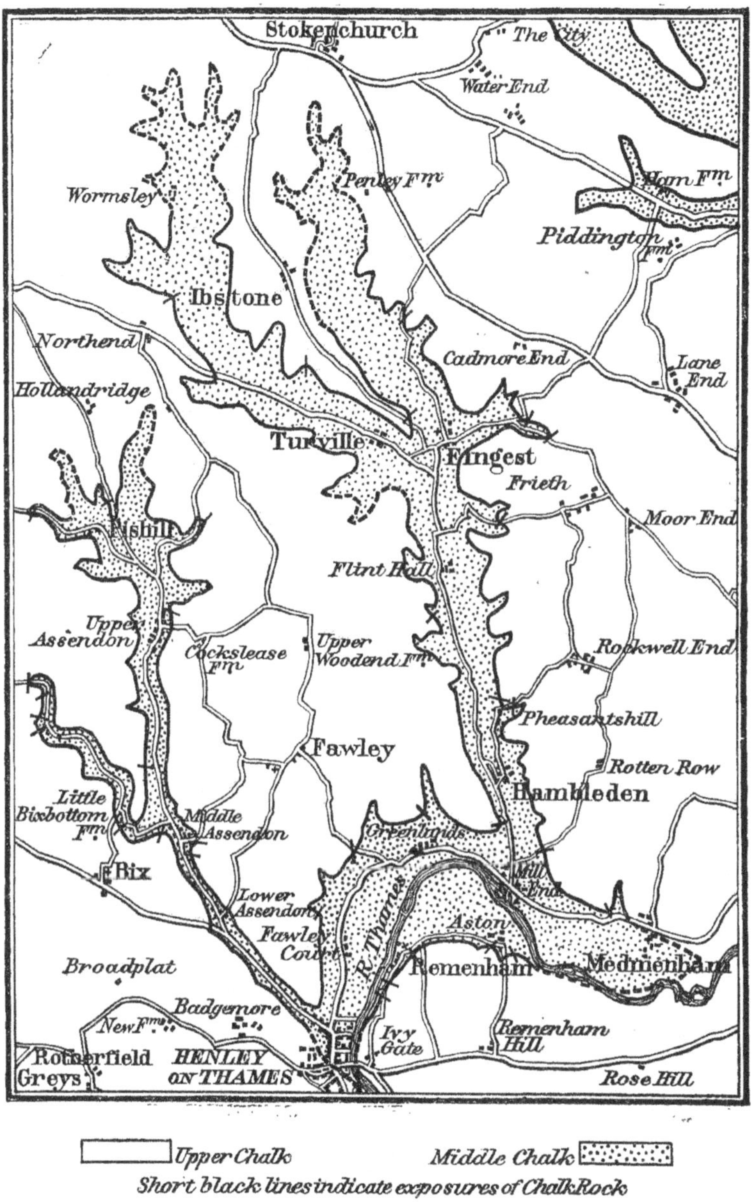


*Fig. 1—Map Showing the Outcrop of the Chalk Rock north of Henley-on-Thames.*

# CHAPTER I.

## INTRODUCTION.

By A. J. Jukes-Browne.

The area included within Sheet 254 of the Geological Survey map is about 216 square miles. It comprises parts of Berkshire and Oxfordshire, with a small portion of Buckinghamshire. The chief towns are Henley, Wallingford, and Watlington, but in former days the Roman station of Dorocina (now Dorchester), and the British settlement on Sinodun Hill were places of importance.

### *Physical Features.*

The area under description exhibits two very different types of scenery, all the western part of it consisting of low-lying land through which rivers and streams meander in sinuous courses, while the eastern part presents a network of hills and ridges separated by deep and generally dry valleys. These two areas are divided from one another by the slope which culminates in the long range of the Chiltern Hills, this range crossing the district from S.S.W. to N.N.E., and gradually increasing in height as it is followed from south to north. Thus east of Ipsden the highest point is 651 ft.; at Nuffield the ground rises to 675 ft., on Brightwell Hill to 753 ft., and on Christmas Common east of Watlington to 790 feet. The Downs above Shirburn and Lewknor rise to heights of 825 and 837 ft., and Crowell Hill, near the northern border of the area, is 806 ft.

The principal river is the Thames, a portion of which traverses the western part of the area, and after taking a southward bend appears again in the south-east corner of the map. The Thame, coming from the north, joins the Thames or Isis at Dorchester, and other small streams flowing outward from the foot of the Chiltern Hills pour their waters either into the Thame or the Thames above Streatley. The drainage from the eastern slopes of the Chiltern Hills is gathered into valleys, the chief of which open into the northerly bend of the Thames near Henley.

### *Formations.*

The formations which enter into the structure of the district belong to four distinct geological systems; these in descending order of superposition are—

4. The Pleistocene;
3. The Eocene;
2. The Cretaceous;
1. The Jurassic.

Below is a tabular view of the subdivisions which have been made in these rock-groups, and are indicated by colours on the published geological map; the maximum thickness which each attains in the district is also given.

*Table of Geological Formations.*

| Groups. | Sub-divisions. | Thickness in Feet. |
|---|---|---|
| Recent and Pleistocene | Alluvium - | — |
| | Valley Gravel - - - | 30 |
| | Plateau Gravel - - | 20 |
| | Pebble Gravel - | 15 |
| | Clay-with-flints - - | — |
| Lower Eocene - | London Clay - | 50 |
| | Reading Beds - | 60 to 70 |
| Upper Cretaceous - | Upper Chalk - - | 260 to 300 |
| | Middle Chalk - - - | 200 |
| | Lower Chalk - - - - - | 200 to 220 |
| | Selbornian { Upper Greensand - | 100 |
| | Selbornian { Gault - | 200 to 220 |
| Lower Cretaceous - | Lower Greensand - - | 80 |
| Upper Jurassic - | Portland Beds - - | 351 |
| | Kimeridge Clay - - - | 110 |
| | *Corallian*[1] - - - | 60 to 80 |
| | *Oxford Clay*[1] - - | 450 |

[1] Not exposed at the surface.

It should be mentioned that not one of these rock-groups is here complete, each being separated from that above by a break and unconformity. Thus between the Jurassic and Cretaceous series there is a gap represented in the south of England by the Purbeck Beds, the Wealden, and the older part of the Lower Greensand. Between Lower and Upper Cretaceous there is a smaller break. Between the Cretaceous and the Eocene is another gap, for the greater part of the Upper Chalk is wanting, as well the Thanet Sands at the base of the Eocene. Lastly, above the London Clay, all the rest of the Eocene and the Oligocene series is wanting.

The physical features of the country are in great part due to the difference in the lithological composition of the rocks and to the manner in which the strata are arranged. These crop out to the surface in regular order; the general dip of the beds being to the E.S.E., so that the strike or direction of outcrop crosses the area from S.S.W. to N.N.E. In consequence of this dip the lowest and oldest beds are found in the N.W. corner of the map, and anyone walking from that point in a southeasterly direction will come to and pass over beds that are successively higher in the geological scale.

The existence of the Chiltern Hills is another direct consequence of the dip of the strata, this range being part of the great escarpment of the chalk which runs across England, and its development is due to the greater resistance which the Chalk presents to detritive agents. The soft sands and clays of the country to the west of it have been worn away by the action of rain and streams more rapidly than the substance of the Chalk which absorbs a large portion of the rain falling on its surface, and allows it to pass away underground. Some of this water is thrown out in springs on the outer slope of the escarpment, but much of it flows eastward under the London basin.

The relation of towns and villages to the natural sources of water supply is well shown in the tract of country with which we are dealing. Almost all the village settlements in the western half of the area are situated either along the courses of the Thames and its tributary streams, or else near points where strong springs issue from the Chalk. Thus there is a succession of villages along the outcrop of the Lower Chalk, the sites of North Stoke, Mongewell, Crowmarsh, Ewelme, Britwell Salome, Cuxham, Watlington, Shirburn, Lewknor, Aston Rowant, and Crowell being all determined by the issue of springs which afforded a good supply of water.

Within the line of the Chiltern Hills villages are sometimes placed in the valleys and sometimes on the ridges, according as local conditions afforded some supply of water and local effort sunk wells in pursuit of it.

### *Faults.*

Few faults exist in the area under description, and none of them causes any great displacement. South of Moulsford, the outcrop of the Lower Chalk is terminated by a fault which strikes from west to east, and carries the Middle Chalk eastward to the alluvium of the Thames. Its throw is probably from 70 to 80 ft.

Another and longer fault runs in a curving line from Berwick Prior through Cuxham and Pyrton to the neighbourhood of Lewknor. This brings the malmstone of the Upper Greensand against the Lower Chalk, but the displacement cannot be more than 40 or 50 ft.

---

# CHAPTER II.

# JURASSIC.

By H. B. Woodward.

## Oxford Clay.

This is the oldest formation proved to occur in the area. It consists mainly of clay and shale with occasional cement-stones or septaria; but the lower portion, within 50 ft. of the base includes beds of sand and sandstone with, at the base, about 10 ft. of clay. These strata are grouped as the Kellaways Beds. The full thickness of the formation is about 450 ft.; but it is nowhere exposed at the surface.

## Corallian.

This is a variable group of limestones with clayey and sandy layers in the upper part, and of sand and sandstones in the lower. The total thickness may be as much as 80 ft., judging by a boring at Shillingford (see below). It is not exposed in the area.

## Kimeridge Clay.

This formation is rather more than 100 ft. thick, and consists of dark shaly clay with earthy limestone nodules and septaria. The upper beds merge gradually into the Lower Portland Beds. Phosphatic nodules occur at the base, as noted in the following record of a boring.

At Shillingford, to the north of Wallingford in Berkshire, a boring was made in 1885 by Messrs. Isler and Company. The section has been described by Mr. Jukes-Browne, who records the strata as follows[1] :—

| Boring at Shillingford. | | Thickness. | | Depth. | |
|---|---|---|---|---|---|
| | | Ft. | In. | Ft. | In. |
| | Soil, &c. - - | 4 | 0 | — | |
| | Sand and gravel | 15 | 0 | 19 | 0 |
| Gault - | - - - | 144 | 0 | 163 | 0 |
| Lower Greensand | - - . . . | 25 | 0 | 188 | 0 |
| Kimeridge Clay. 111 ft. 6 in. | Stiff grey clay and stones (? septaria) | 20 | 0 | 208 | 0 |
| | Black clay, with shells and stones, and phosphatic nodules at 296 ft. | 91 | 6 | 299 | 6 |
| Upper Corallian. 44 ft. 6 in. | Rock - - - - - | 12 | 9 | 312 | 3 |
| | Stony clay and shells (yellow gritty loam) - - | 5 | 8 | 317 | 11 |
| | Rock, with some layers of clay - | 9 | 11 | 327 | 10 |
| | Stony clay, with layers of rock - | 8 | 10 | 336 | 8 |
| | Rock, with shell fragments (*Ostrea*) - | 7 | 7 | 344 | 3 |
| Lower Corallian. 35 ft. 3 in. | Sand and stone - - - - | 7 | 4 | 351 | 7 |
| | Alternations of hard grey rock and clay - - | 10 | 4 | 361 | 11 |
| | Dark gritty clay and shells | 10 | 1 | 372 | 0 |
| | Hard grey limestone | 2 | 6 | 374 | 6 |
| | Sand and stone with water | 5 | 0 | 379 | 6 |
| Oxford Clay | Blue clay, soft and slightly mottled - | 4 | 6 | 384 | 0 |

[1] *Midland Naturalist*, vol. xiv, p. 201.

The water obtained at the base of the Corallian rose to within 60 ft. of the surface, and although palatable it did not yield a sufficient supply. Curiously enough that obtained from the Lower Greensand above was saline, containing 98 grains per gallon, 54 of which were chloride of sodium.

At Culham, to the south-east of Abingdon, there is evidence of the presence of the passage-beds from Kimeridge Clay to Portland Beds, for John Phillips in 1860 noted the occurrence there of fossils like those of Hartwell, near Aylesbury.

## Portland Beds.

The Portland Beds of the district consist of the following strata:—

| | |
|---|---|
| Upper Portland Beds | Sand with hard ferruginous bands.<br>Whitish limestones.<br>Clays and greenish sands.<br>Glauconitic limestone with lydites. |
| Lower Portland Beds | Blue and brown clay with lydites.<br>Yellow and greenish-grey sands with concretionary masses of sandstone. |

The upper beds yield the following fossils:—

Perisphinctes boloniensis (*de Lor.*)
Cerithium portlandicum *J. de C. Sow.*
Natica elegans *J. de C. Sow.*
Cardium dissimile *J. de C. Sow.*
Lucina portlandica *J. de C. Sow.*
Ostrea expansa *J. Sow.*
Pecten lamellosus *J. Sow.*
Trigonia gibbosa *J. Sow.*

Among fossils from the lower beds the following have been recorded:—

Perisphinctes giganteus (*J. Sow.*)
Perisphinctes pectinatus (*Phill.*)
Lucina portlandica *J. de C. Sow.*
Ostrea expansa *J. Sow.*
Pecten nitescens *Phill.*
Perna mytiloides *Lam.*
Pholadomya rustica *Ag.*
Pleuromya tellina *Ag.*
Pinna.
Trigonia pellati *de Lor.*

The sands and limestones of the Portland Beds occur over a small area to the north-east of Stadhampton, and north-west of Warpsgrove; but no noteworthy sections have been observed.[1]

In his notes on the section at Culham brickyard to the south-west of Nuneham Courtenay, and beyond the limits of our map, Phillips described a bed of green sand, 9 ft. thick, below the Lower Greensand and Gault, and above the Kimeridge Clay, from which there were evidences of "an easy passage upwards from clay to sand." He obtained from this sand a few examples of the Ammonite like *A. polyplocus*, *Cardium striatulum*, *Thracia depressa*, *Pecten arcuatus*, *Corbula*, and wood; and he suggested that the bed might be "perhaps the first stage of a change towards Portland series."[2]

At a later date, when the first Geological Survey was in progress, Prof. Hull found some fossils in "the road cutting at

[1] General descriptions of the Jurassic Rocks, with lists of fossils, have been published in the Memoir on 'The Jurassic Rocks of Britain,' vol. v, 'Middle and Upper Oolitic Rocks of England (Yorkshire excepted),' by H. B. Woodward, 1895, pp. 167, 169, 217.

[2] *Quart. Journ. Geol. Soc.*, xvi 1860 p. 310.

Toot Baldon," north of March Baldon, and of the area shown on the map. The discovery is thus mentioned[1]:—

"At this place marine shells (*Ammonites Deshayesii* and *Terebratula sella*) of the Lower Greensand period were found, which are extremely valuable in determining the condition under which these beds were deposited."

The late J. H. Blake, while mapping this area found some blocks of more or less cherty sandstone with casts of fossils, in a field 700 yds. east of March Baldon Green. The fossils were poorly preserved, but they were examined in the Palæontological Department of the Survey by Mr. E. T. Newton, who reported that:

"Although the specimens can scarcely be taken to prove the strata to be of Lower Greensand age, yet their general facies points to this conclusion."[2]

It has been noted by Mr. Lamplugh that cherty beds are not known to occur elsewhere in the Lower Greensand of this district, and that cherty blocks since found on the slopes of Shotover Hill prove to be altered Portlandian rocks.[3]

During his examination of the strata between Culham and Toot Baldon he had observed a change in the lithological character and colour of the beds northwards from the neighbourhood of March Baldon. There they consist of buff or greyish sands of finer grain than the coarse and pebbly Lower Greensand of Culham and Clifton Hampden, and of great portion of Nuneham Park. Mr. Osborne White has likewise noted these differences.

Meanwhile the Lower Cretaceous strata in this area had been described in 1898 as the "Toot Baldon Beds," by Dr. A. Morley Davies; although he admitted the possibility that there might be a trace of the lower sandy beds of the Portlandian at Toot Baldon.[4]

During the early part of the year 1907 Dr. Davies very kindly sent to the Museum at Jermyn Street, for determination, some fossils which he had collected in 1903 from sandy strata in the lane-cutting south of Oldhouse Farm, Toot Baldon. The specimens include a clavellate form of *Trigonia* not sufficiently perfect to be identified specifically; also *Cyprina*?, *Exogyra thurmanni*? (Etallon), and *Protocardia*? As the evidence supplied by these fossils was not regarded as conclusive, it was decided that Mr. John Pringle should visit the area; and although he was unsuccessful at the locality where Mr. Davies obtained his specimens, other fossils of identical matrix were obtained in the neighbourhood, some of them of similar species to those in the Lower Portland Beds at Shotover Hill, and preserved in rock of similar lithological character. Mr. Pringle

1 'The Geology of Parts of Oxfordshire and Berkshire' (*Mem. Geol. Surv.*), 1861, p. 15.

2 'Summary of Progress' (*Mem. Geol. Surv.*) for 1900, p. 120. The specimens referred to are not preserved in the Survey collection.

3 The Cherty layers mentioned as occurring in the Shotover Iron-sand and Ochre Series (see Phillips, Geol. Oxford, pp. 414, 415) were evidently at the base of the series and probably Portlandian.

4 *Proc. Geol. Assoc.*, vol. xvi, p. 45.

ascertained that in the fields east of March Baldon and at Baldon Row, highly fossiliferous "doggers" of sandstone are frequently met with at a depth of 6 to 8 ft. from the surface. He further noted that the sands with Portlandian fossils are fine grained and quite distinct from those of the Lower Greensand; thus confirming the suggestions made by Mr. Lamplugh.

A line separating the Portland Beds and Lower Greensand has consequently been drawn in accordance with the evidence that could be gathered of the distribution of the different sandy deposits between the definite areas of Lower Greensand on the south, and of Portland Beds on the north. This line was inserted on the new Oxford map, but it was not shown on the map now under description, Sheet 254, which was previously printed.

The following fossils were collected from the Portlandian Beds by Mr. Pringle:—

Garden of house 440 yds. W.S.W. of St. Lawrence Church, Toot Baldon. [Pr. 3978-3984].

| | |
|---|---|
| Astarte rugosa? (*J. de C. Sow.*) | *cf.* Lucina portlandica *J. de C. Sow.* |
| Corbicella? | Pecten *cf.* morini *de Lor.* |

Field 900 yds. N.E. of Post Office, March Baldon. [Pr. 3989-4000]. (Portlandian).

| | |
|---|---|
| Astarte rugosa (*J. de C. Sow.*) | Pecten morini *de Lor.* |
| Cyprina? | ,, nitescens *Phill.* |
| Leda? | ,, sp. |
| Ostrea sp. | Perna sp. |

It should be mentioned that Mr. Pringle noted the strong resemblance of the sandy Portland bed seen at the southern end of the Culham pit, and previously mentioned, to the material composing the doggers of Portlandian sandstone met with at Toot Baldon; and he obtained from the Culham sand-bed a doubtful *Serpula* and a specimen near to *Isodonta longior* (Blake).[1]

[1] These fossils [Pr. 4004-4007] are preserved in the Museum at Jermyn Street, See also 'Geology of the Country around Oxford' (*Mem. Geol. Surv.*), 1908. In the Memoir on 'The Geology of the Country around Oxford' (1908), p. 51, ragments of an ammonite from Portland Beds at Shotover Hill were named *Perisphinctes plicatilis?* (J. Sow.). These should not have been identified so nearly with this species, but should have been merely recorded as "*Perisphinctes sp.*"
F. L. K.

# CHAPTER III.

## LOWER CRETACEOUS.

### Lower Greensand.

The following passages by Mr. G. W. Lamplugh are reproduced from his description of a portion of the Lower Cretaceous rocks, in the Memoir on 'The Geology of the Country around Oxford,' 1908 (pp. 77, 78):—

A broad tract of Lower Greensand was shown in the original Geological Survey map as covering Nuneham Park, and extending northward to Toot Baldon. In the new series map Sheet 254, which was colour-printed in 1905, Lower Greensand was represented over the area of March Baldon to the northern margin of the map at Baldon Row, as surveyed by J. H. Blake.

The mapping of this tract has been admitted, by all who have examined the ground, to present many difficulties, but as elsewhere noted (see p. 6), evidence has now been obtained to show that Portlandian beds occur over this part of the area, probably from the northern confines of Nuneham Park over the uplands of March Baldon and Toot Baldon, and eastwards towards Chislehampton.

The evidence of fossils indicates conclusively that Lower Greensand occurs at Culham and Clifton Hampden to the south of Nuneham Park, and that Portlandian occurs at March Baldon, and also at Toot Baldon, just beyond the limits of our present area, on the north. The lithological evidence, as already pointed out, shows that the Lower Greensand locally consists of coarse and ferruginous pebbly and quartzose sand; the Portlandian of fine-grained sands and sandstone.

That the Lower Greensand is subject to rapid changes in thickness is shown in the brickyard section at Culham, where it is diminished to a few inches, though seen to a depth of 10 ft. in a sand-pit only 150 yds. farther west. In this sand-pit the strata consist of coarse gritty ferruginous sand with small pebbles, shown to a depth of 12 ft.; and similar material, still more pebbly, is exposed in a low cliff on the Thames bank at Clifton Hampden. The beds in this region are locally indurated into rocky layers.

In a MS. descriptive of the neighbourhood, written by Daniel Sharpe,[1] there is a statement that organic remains had been obtained from the sand "in a low bank by the side of the road which slopes from a turnpike in the Abingdon road down towards the ferry at Clifton: they are the following species in the state of ferruginous casts occasionally preserving the outer surface:—*Exogyra conica*, *Terebratula ovata*, *Rhynchonella octoplicata*,

1 Preserved in the Library of the Geological Survey and Museum.

and *Rhyn. Hampdeni*, a new species found also in the Upper Greensand of Warminster."

There is no doubt that these are marine deposits, as marine fossils have been found in the sandy layer between the Gault and the Kimeridge Clay in the Culham brickyard section. Phillips[1] recorded a *Pecten* and an *Exogyra* from this bed; and the present writer, in a brief examination of the pit, in the summer of 1902, found the casts of several shells, including a small oyster and a *Trochus*-like univalve.

Mr. Jukes-Browne, however, considers the sandy bed of the brickyard to be the basement-bed of the Gault, which is supposed to cut out the Lower Greensand at this point[2]. The indurated portion of the sands in the adjacent sand-pit contain markings that appear to be the casts of fossils, and recently a large specimen of *Exogyra sinuata* J. Sow has been found in a bed of coarse laminated sandstone at Culham.[3] Some fossils have lately been collected for the Geological Survey, by Mr. J. Pringle, in the sand-pit west of Culham brickyard: they include the following:—

Radiopora sp.
Exogyra sinuata *J. Sow.*
*cf.* Venus orbigniana *Forbes.*

G. W. L.

The difference of opinion referred to in the foregoing paragraph, however, is more apparent than real, as the zone classed by Mr. Jukes-Browne as belonging to the base of the Gault is taken by Mr. Lamplugh to be the top bed of the Lower Greensand. The zone in question—that of *Douvilleiceras mammillatum*—is probably represented by the gritty band at the base of the Gault in the Culham section, as fossils recently obtained by Mr. Osborne White have indicated. (See p. 14.)

J. H. Blake noted exposures of Lower Greensand north of Hill Farm and by Warren Farm, north of Culham; also eastwards to the railway. These comprise pebbly sands of like character to the Lower Greensand before mentioned.

By Rough Lodge, south-east of Nuneham Park, Blake, in 1896, recorded that a brickyard had been opened in 1852, and had since been abandoned. He noted 4 ft. of coarse quartzose sand resting on 4 ft. of thinly-bedded, coarse, ferruginous sandy rock, locally known as "Roach." Dark blue clay was said to have been dug, and clay also was said to have been obtained from Clifton Heath, where an outlier of Gault has been mapped.

Coarse, pebbly and quartzose sand was observed by Blake in Lock Wood, on the western side of Nuneham Park. There the material known as "Roach" has been used for making paths. Similar sand was observed by him to the east and north-east of the Mansion in Nuneham Park. Here the difficulties of

1 *Quart. Journ. Geol. Soc.*, xvi (1860), p. 309; and 'Geology of Oxford,' p. 428.
2 'The Cretaceous Rocks of Britain' (*Mem. Geol. Surv.*) i, 1900, p. 268. See also H. J. O. White, *Proc. Geol. Assoc.*, xviii, 1903, p. 300.
3 C. G. E. Dawkins, *Geol. Mag.*, 1906, p. 94.

distinguishing Lower Greensand from Portlandian are complicated by the presence of patches of Drift sand and gravel, and the absence of good sections. H. B. W.

The following notes are by Mr. Jukes-Browne:—

In the small riverside cliff at Clifton Hampden, the height of which is about 20 ft., the rock exposed is a sandstone, consisting of coarse brown quartz sand, some of the grains being large enough to be called pebbles, but few are bigger than peas, the mass is bound together by a network of ferruginous veins or seams, and is very hard in the lower part, but the upper portion is disintegrated. The grains and pebbles are nearly all of quartz, but there are a few of a dull yellowish stone which is probably an indurated Jurassic rock (perhaps Corallian).

At the brick-kilns on Clifton Heath near the south-east corner of Black Wood, pits have been dug to obtain sand for brickmaking purposes, and one of these exposed the following beds in 1887:—

| | Ft. | In. |
|---|---|---|
| Coarse pebbly sand or small gravel, like that of Clifden Hampden, but loose - - - | 3 | 0 |
| Passing down into coarse angular yellow and brown quartz sand | 1 | 6 |
| Pebbly sand like that above compacted by iron into a ferruginous rock - - - - - - | 0 | 6 |
| Coarse yellow sand passing down into fine sand, seen to - - - - - - | 4 | 0 |

I was informed by the workmen that the fine yellow sand became grey below and harder; a heap of such hard grey sand lay near the edge of the pit, and was said to have come from about 8 or 9 ft. below the ferruginous band.

A. J. J.-B.

# CHAPTER IV.

## UPPER CRETACEOUS.

### SELBORNIAN (GAULT AND UPPER GREENSAND).

BY A. J. JUKES-BROWNE.

The *Selbornian*, as indicated on p. 2, includes the Gault and Upper Greensand, and so long as these old names are used merely to designate special lithological facies of the formation there is no objection to their employment. As, however, in some localities the whole formation consists of clay and marl (Gault), while in others it is almost entirely composed of sand (Upper Greensand and Blackdown Beds), it is clear that a name for the group as a whole was wanted in order to prevent misconception.

Moreover, there are large areas over which another kind of rock material forms a prominent member of the formation; this is *malm* or *malmstone*, a fine-grained siliceous rock, which cannot be properly called a sandstone, though the upper part of it often passes laterally into a fine micaceous sand or sandstone. Consequently it often happens that the Selbornian comprises three or four different kinds of rock-material in regular succession, and this is the case in Berkshire and Oxfordshire, where the succession may be stated as follows:—

| | | | Ft. | In. |
|---|---|---|---|---|
| Upper Greensand | Glauconitic marl and sand- | 20 ft. to | 25 | 0 |
| | Malmstone - | 50 ft. to | 80 | 0 |
| Gault - - | Silty Marls - | | 200 | 0 |
| | Grey clays - - - - | | | |

The clays and marls have been mapped by the Geological Survey as "Gault" and the beds above as "Upper Greensand," but the zonal arrangement of the beds is probably as follows:—

| | |
|---|---|
| Zone of *Pecten asper* - - | Glauconitic marl and sand. |
| Zone of *Schloenbachia rostrata* | Malmstone<br>Silty marls. |
| Zones of *Hoplites lautus* and *H. interruptus* - | Grey clays. |

It should be mentioned, however, that very little palæontological evidence has been obtained in the area under description.

The existence of malmstone in this part of England was first recognised in 1885, by Dr. G. J. Hinde.[1] It is a whitish rock, with very little true sand in its composition, consisting largely of sponge spicules and globular colloid silica, with some mica and a certain proportion of marly or calcareous matter. In some places it forms a massive grey rock, but more often there are alternating beds of compact malmstone and fine siliceous marl.

1 'On Beds of Sponge Remains in the Lower and Upper Greensand of the South of England.' *Phil. Trans.*, 1885, p. 416.

West of Wallingford its thickness may be over 80 ft., but it gradually thins eastward, becoming at the same time more and more interstratified with marl till near Chinnor its total thickness is probably not more than 40 ft.

Fossils are not abundant in the malmstone, *Hoplites auritus* (J. Sow.), *Aucellina gryphæoides* (J. de C. Sow.), and *Pecten* (*Syncyclonema*) *orbicularis* J. Sow. being those most usually seen.

The malmstone in its central part is a light grey siliceous and micaceous stone, which dries nearly white, and might then be mistaken for chalk when viewed from a little distance. Its specific gravity is small, and microscopical examination has proved that it is largely composed of organic silica derived from the dissolution of siliceous sponges.[1] Much of it is similar to the malmstone of Farnham, which Dr. Hinde describes as consisting almost entirely of organic silica either in the shape of detached spicules or in the form of minute globules. He also examined specimens from Wallingford, which he thus describes:—"The rock is hard, and calcite is only occasionally present in it, but there is a greater relative amount of quartz-sand and other mechanically derived constituents than in the malm of the Wealden district. The sponge-remains are either as empty casts or of amorphous silica, and the siliceous matrix is partly in the globular form" (*op. cit.*, p. 416).

Specimens from Cuxham, near Watlington, which I sent to Dr. Hinde were found to be similar, and in writing to me he says, "there are plenty of Foraminifera in all the specimens, principally *Globigerina*, *Textularia* and *Rotalia*. In one of the specimens minute coccoliths are present."

When a hand specimen of this malmstone is examined a number of glistening particles are seen, most of which are probably scales of mica, but some, as Mr. W. Hill has suggested to me, may be the broken ends of sponge spicules. The quartz-sand referred to by Dr. Hinde is in the condition of fine silt, and is not visible to the eye; minute grains of glauconite are also abundant in the rock.

The above description applies only to the central portion of the malmstone mass, both the lower and upper parts becoming more earthy or marly, and containing a less proportion of organic silica.

Above the malmstone a band of marl containing large grains of glauconite has been seen at several places, and is probably continuous for some distance. It may be from 6 to 9 ft. in thickness. The series is completed by an equally thin band of glauconitic sand, consisting entirely of quartz and glauconite grains, but no fossils have been found in these beds either near Wallingford or in Oxfordshire.

[1] See Hinde, *Phil. Trans.*, 1885, p. 403.

## *Stratigraphical Details.*

### GAULT.

The basset surface of the Gault in the area comprised within Sheet 254 is largely covered by gravel, and even in those tracts where it is at or near the surface there are very few exposures.

At Didcot, just outside the western margin of the map, the uppermost marls are well exposed in the railway cutting near the station. At the bottom of this cutting there is tough micaceous marly clay which passes up into soft grey micaceous marl, containing *Aucellina gryphæoides* (J. de C. Sow.); small phosphatic nodules are scattered throughout the marls. In the old Geological Survey Map (Sheet 13) these marls were included in the Upper Greensand, and similar marls, with occasional floors of malmstone, are exposed in the first cutting on the Didcot and Newbury line. These are clearly the local passage beds from Upper Gault marls to the "Upper Greensand" or Malmstone group, and the line drawn on the map is of no value except as a rough boundary line between the two local lithological facies of the formation.

The same marly beds can be seen at the north-western end of the long railway cutting S.W. of North Moreton, and again in the roadway up the hill south of Little Wittenham.

An interesting section is exposed in the brickyard at Culham, and the section seen at the eastern end of the pit in 1887 was as follows:—

| | | Ft. | In. |
|---|---|---|---|
| Gault | 4. Light grey laminated clays, very micaceous in the lower part, *Hoplites interruptus*, etc. - - | 16 | 0 |
| | 3. Hard sandy clay full of small quartz pebbles, with some phosphatic nodules and casts of fossils | 2 | 0 |
| Kim. Clay. | 2. Bright brown sandy ferruginous limestone | 0 | 6 |
| | 1. Dark grey clay - - - | 16 | 0 |

Prof. Phillips has recorded the following fossils from this locality[1] :—

Ammonites dentatus [Hoplites interruptus (*Brug.*)]
Ammontes lautus
Belemnites minimus
Dentalium, probably D. decussatum
Rostellaria
Solarium conoideum
Inoceramus concentricus (large)
Nucula pectinata
Pecten quinquesulcatus [P. quinquecostatus *J. Sow.*]
Plicatula pectinoides
Cyclocyathus Fittoni
Balanus
Coniferous wood.

Mr. Osborne White described the section[2] visible at this place in 1903 and his account agrees closely with that above given. The face had doubtless been cut further back and his measurements were probably taken more to the west than mine, for the exposure described by him includes pockets of the surface gravel, the thickness of the Gault clays being about 20 ft. and the pebbly basement bed having diminished to 13 in. He

1 *Quart. Journ. Geol. Soc.*, vol. xvi, p. 309.
2 *Proc. Geol. Assoc.*, vol. xviii, p. 303.

remarked that the Gault here worked "formed part of a north-westerly projecting tongue of the main mass underlying the low ground on the south side of the Thames, and had evidently been deposited in a trough-like hollow of which the sides consisted of Lower Greensand and the bottom in part of Kimeridge Clay."

The following fossils have recently been obtained from the Culham pit by Mr. Osborne White:—

*Bed 3.*—
- Pentacrinus sp.
- Serpula sp.
- Inoceramus concentricus *Park.*
- Lima (Plagiostoma) globosa (*J. de C. Sow.*)
- Lucina sp.
- Pecten (Syncyclonema) orbicularis *J. Sow.*
- " (Neithea) quinquecostatus ? *J. Sow.*
- " (Chlamys) robinaldinus *d'Orb.*
- Pteria sp.
- Fragments of two ammonites, ill-preserved. These are probably *Cleoniceras beudanti* (Brongn.) or an allied form, and *Douvilleiceras mammillatum* (Schloth.).

*Bed 4.*—
- Hoplites interruptus (*Brug.*)
- Belemnites minimus *Lister*
- Inoceramus concentricus *Park.*

The above were named in the Palæontological Department.

The only other noteworthy exposure of Gault is at the Clare Brickyard, marked "Brick Works" on the old 1 in. map, about a mile N.W. of Clare. This was visited by Mr. Blake in 1898, who contributes the following note; the clay is of a light bluish-grey colour with a little "race," a few ferruginous concretions and small phosphatic nodules; the face exposed was only 3 ft. deep, but other portions in previous years had been worked to a depth of 21 ft., and according to the foreman many fossils had been found. From this brickyard Mr. Blake obtained a specimen of *Schloenbachia rostrata* (J. Sow.).

Mr. J. H. Blake noted in 1900 that "The hill to the north-west of Chislehampton has been proved by a boring 60 ft. in depth, made this year near the top of the sloping ground opposite Church Farm, to consist of bluish-grey clay, blackish in places. It is apparently Gault overlying Kimeridge Clay."[1]

A small area of Gault was mapped by J. H. Blake on Pebble Hill, Toot Baldon—a hill named from a house called "The Pebbles" (now the Roebuck Inn), in the wall of which nine pebbles are inserted in the form of a diamond.

There are tracts of clayey ground on March Baldon Green, and around Windmill Hill in Nuneham Park. These may be Gault.

## Upper Greensand.

It will be convenient to indicate first the principal exposures in the malmstone, which forms the main mass of the Upper Greensand, and then to trace the course of the narrow band of glauconitic marl and sand which lies between the malmstone and the Lower Chalk.

Where the beds of malmstone begin to come in there is always a marked rise in the ground, in other words the outcrop of the

[1] 'Summary of Progress' for 1900 (*Mem. Geol. Surv.*), p. 120.

malmstone forms a ridge with a fairly steep slope or escarpment to the north and north-west, and a gentle slope in the direction of the dip.

The malmstone tract enters the area of this map between Didcot Newtown and East Hagbourne. It has been dug by the roadside south of Didcot Church and there are good sections of it in the cuttings on the line to Newbury south-east of Didcot. The beds seen in the first of these have been mentioned on p. 13; in the next cutting at the north-west end there is at he base a dark sandy clay (Upper Gault), like that at Didcot, succeeded by grey sandy marl which passes up into soft light grey malm with bands of hard calcareous malmstone; these beds occupy a syncline at the base of which a nearly black sandy marl is seen, apparently replacing the upper part of the grey marl. The beds then rise to the south-east and the first bridge passes over the centre of a low anticlinal curve, this being well marked by a bed of hard siliceous stone which has a yellowish staining. South-east of this bridge there is a second syncline, and then a rolling dip which brings in higher beds of light grey siliceous stone near the second bridge.

At Haddon Hill, on the road to Wallingford, there is a small pit in malmstone showing a hard yellowish bed like that of the cutting dipping at 3° westward. There is another pit about 6 furlongs further west by the road.

Sinodun and Brightwell Hills consist mainly of malmstone, exposures of which can be seen by the roadside south-west of Little Wittenham and in a shallow cutting south of Castle Hill.

Behind the farm buildings south of Brightwell Hill a bank shows 6 ft. of soft grey sandy marl with lumps of hard calcareous stone, and a small pit below is dug in massive grey siliceous and micaceous stone of a brownish-grey colour when wet, but drying to a light chalky-looking rock.

At Satwell spring-head there is a soft greenish-grey stone overlain by soft grey sandy marl, and in the lane south of the Bell Inn soft micaceous malmstone can be seen in the bank.

The best exposure in this district, however, is in the road-cutting near Shillingford Bridge, a considerable thickness of light grey malmstone is here seen, drying so white that it looks like chalk until closely examined; below it is grey sandy marl; it was doubtless from this exposure that the specimens examined by Dr. Hinde were taken (see p. 12); the stone has also been quarried in the cliffs behind the Swan Hotel.

On the eastern side of the Thames valley the malmstone beds emerge from beneath the gravels near the village of Roke, and rise into a well-marked ridge which runs thence to the north-east. The malmstone is exposed in a small pit about half a mile N.E. of Roke, and again in a larger pit where the road crosses Hollantide Bottom. From these two quarries were obtained—

Hoplites auritus (*J. Sow.*)
Pecten (Syncyclonema) orbicularis, *J. Sow.*
Aporrhais parkinsoni (*J. de C. Sow.*)
Ostrea sp.
Grammatodon carinatus (*J. Sow.*)
Echinospatagus sp.

The base of the permeable beds is marked by the spring-head N.E. of Berrick, where a layer of hard micaceous stone rests upon soft clunchy marl.

Malmstone has been dug near Whitehouse Farm, and again at the corner of the lane east of Cadwell Farm, where the stone is grey and apparently more calcareous. At Cadwell Spring there is similar stone overlying marly clay, the former containing large *Aucellina gryphœoides* (J. de C. Sow.). Thence the beds pass through Brightwell Park, and are exposed by the side of the road near Brightwell Rectory; a strong spring rises in the Rectory grounds, and the water is led beneath the road and the watercourse has been widened into a watercress-bed.

At Cuxham soft rotten malmstone can be seen in the watercourse below the corn-mill, and good hard whitish stone is exposed at the entrance of the field road to Pyrton Heath Farm.

Between Cuxham and Easington the malmstone ridge is divided into two parts by a longitudinal valley, and the best sections occur on the outer ridge. Light grey malm of light specific gravity is exposed in a pit two furlongs S.E. of Easington and in the road cutting just above.

The road cutting down Clare Hill gives a nearly complete section of this division. At the top near the guide-post there is a tough calcareous marl with a few scattered green grains, the next thing seen is hard grey malmstone, and there is at least 40 ft. of this; opposite the bench mark of 330 ft. a layer of malmstone is seen overlying a foot or two of softer micaceous marl; no stone was seen below this, the lowest water-hole showing 3 or 4 ft. of tough grey micaceous and calcareous marl with *Hoplites auritus*, which might be classed as Malmstone marl or Upper Gault marl.

Near Clare South Farm there is a pit in siliceous malmstone, which does not seem to contain any fossils except *Aucellina gryphœoides* and *Pecten orbicularis*. Another pit by the roadside, just south of Stoke Talmage, shows about 8 ft. of similar malmstone in regular beds, which break up into small angular blocks and pass up into a marly soil.

The next exposure is at Wheatfield, on the south side of the angle in the road east of the church. Here a small quarry shows the following beds:—

| | Ft. | In. |
|---|---|---|
| Soil and weathered grey marl - | 2 | 0 |
| Massive malmstone rock - - | 3 | 0 |
| Tough grey micaceous marl, becoming rather softer below, seen for - | 8 | 0 |

At Adwell there are two fine spring-heads, the water issuing from the malmstone, but no open sections were seen. Beyond Adwell the stone forms a promontory, along which runs the main London and Oxford road and the stone is exposed in the road-cutting less than a mile S.E. of Tetsworth. Thence the feature can be followed to the springs by Copcourt Farm and Postcombe.

At Chinnor by the road to Thame, about half a mile north-west of the church a well was sunk in 1885, and the beds passed through were said by the workmen to be as follows:—

| | Ft. | In. |
|---|---|---|
| Soft grey marl - - - - | 10 | 0 |
| Ragstone in rocky beds separated by layers of grey marl, the lowest being a bed of rock with water at its base - - - | 27 | 0 |

Much of the material brought up lay on the ground, and the stone was seen to be an impure calcareous malmstone, some of the lower beds containing fossils in plenty, *Hoplites auritus*

(J. Sow.), *Desmoceras planulatum* (J. de C. Sow.), and *Aucellina gryphæoides* (J. de C. Sow.), being obtained. From this section it is evident that the malmstone is splitting up and thinning out, soft micaceous marls coming in, and these are found to thicken still more at the expense of the stone beds as the formation is traced through the next Sheet.[1]

### *Glauconitic Beds.*

The outcrop of these forms a narrow band in which exposures are not frequent.

At West Hagbourne (Sheet 253) green sand can be found here and there along the watercourse by the orchard of Manor Farm, but near Upton it passes beneath alluvial deposits, from which it emerges between Blewbury Mill and Sheencroft Farm, the house and buildings of this farmstead standing upon it. A little further eastward, however, it passes below the spread of gravel and alluvium near Aston Tirrold.

It is next seen for a short space along the shallow cutting of the Great Western Railway between Cholsey and South Moreton, but the whole of Cholsey Hill consists of Chalk Marl, and there is no sign of the glauconitic beds on its northern flank, while around Wallingford its outcrop is entirely concealed by the Thames valley gravels.

On the other side of the Thames valley the upper beds of malmstone, and presumably the green sand also, emerge from under the gravels to the east of the hamlet of Roake, but I saw no exposure of the glauconitic beds. Under favourable circumstances green sand might be found in the ditch or watercourse which runs north-eastward to Rumbolds Lane, but this lane is on the Chalk Marl, and to the northward there is malmstone, brought up apparently by a fault, so that the green sand is cut out.

The lowest portion of the glauconitic beds appears to cap the malmstone on the west side of Brightwell Baldwin, for in 1886 I found a soft marl full of green grains in the field-road leading out of the lane 160 yds. south-west of the church; and was also informed by the sexton that such marl "with green specks in it" underlay the greater part of the churchyard. This is on the north side of the supposed line of fault, and no other evidence was obtainable, the ground south of the village being obscured by a great depth of soil and gravelly wash.

At Cuxham no trace of green sand could be found, and from the close proximity of exposures showing malmstone and Chalk Marl respectively, it was inferred that the intervening beds were faulted out. South-east of Cuxham there appears to be a cross fault, and it was surmised that the malmstone and green sand might be brought up on the further side of this fault at the north-east of the "Middle Mill," but no trace of the green sand could be found. The pools and rushy ground north-east of Watlington Mill may possibly indicate the base of the Chalk Marl.

1 See Whitaker, Memoir on Geology of London, vol. i. 1889, p. 52.

The green sand is first seen again at the Model Farm north of Shirburn Castle. Here I was informed by Lord Macclesfield's steward that dark green sand had been found in the orchard, and in the field to the south of it; also that it underlay the gravel on which his house stands, and was passed through in digging the well. Thence the outcrop passes to South Weston, and can be found in the bank of the roadside south of the village; curving round by Nethercroft fishpond it passes below South Weston Church, and runs thence northward through the fields above Adwell Springs; there I proved its outcrop in the dry ditch which runs up from the Springs to the plantation on Adwell Cop.

From Adwell the outcrop was traceable by feature to Aston Rowant, where the dark green sand was found in the bank of the watercourse a quarter of a mile N.N.W. of the church. Thence its course is along low and obscure ground, but it was seen along the watercourse north of Crowell for about 150 yards below the point where the streams from the various springs unite; all these springs rising from the rocky beds in the Chalk Marl, and not from its base, as was erroneously concluded when the old geological map (Sheet 13) and accompanying Memoir were published.

The outcrop of the green sand passes through the northern part of the village of Chinnor.

## OUTLIERS.

Here and there the malmstone plateau supports outliers of the green sand, which are generally capped by patches of Chalk Marl.

Sinodun and Brightwell Hills, north-west of Wallingford, are outliers of this kind, but the outcrop of the glauconitic beds is narrow and its presence can seldom be detected.

On the hill above Sotwell glauconitic marl was seen in the lane leading to the Hill Barn.

The hill east of Easington is capped by a patch of the glauconitic beds from which the Chalk Marl has been removed, but which are covered in part by flint gravel. In the road about two furlongs east of Easington there is yellowish glauconitic sand passing up into hard glauconitic marl with brown patches like soft phosphatic nodules, this appears to be succeeded by soft dark green sand, but it is more likely that the beds have slipped over one another down the slope.

At the cross-roads on Clare Hill soft grey marl full of glauconitic grains is exposed in the bank, and this evidently underlies the green sand and passes down into malmstone.

Gilton Hill, near Wheatfield, is an outlier of the glauconitic beds capped with Chalk Marl, but no exposures were seen.

The manner in which such outliers have been formed by the gradual isolation of a promontory through the recession of spring-heads and the wash of rain is admirably illustrated in the position of Adwell Hill, which may be regarded as a case of the arrested development of an outlier.

# CHAPTER V.

## UPPER CRETACEOUS (*continued*).

### Lower Chalk.

By A. J. Jukes-Browne.

The Lower Chalk, as now defined, includes the beds formerly called the Chalk Marl and the Grey Chalk, together with a certain thickness of the whitish chalk, which overlies the latter. The Lower Chalk is limited by the green sand of the Selbornian group below and the hard nodular chalk of the Melbourn Rock above, and in Berkshire and Oxfordshire its total thickness varies from 180 to 220 ft.

It is divisible into two broad zones—(1) the zone of *Schloenbachia varians,* which corresponds with the Chalk Marl or Totternhoe Marl, as defined by Mr. Whitaker in 1865[1]; and (2) the zone of *Holaster subglobosus,* which has the Totternhoe Stone at its base, and the "Belemnite Marls" or subzone of *Actinocamax plenus* at the summit.

The zone of *Schloenbachia varians* does not consist entirely of soft marly chalk, for it includes many beds of hard grey chalk, which alternate with the bands of softer chalk. One of these hard beds is persistent throughout the counties of Oxford and Bucks, and being underlain by a stiff argillaceous marl, throws out water in a series of strong springs. This bed has been indurated by the infiltration of calcite, and is in some places 4 or 5 ft. thick; its outcrop frequently forms a marked rise in the ground, and is consequently traceable as a feature. As there are good sections of it near Risborough, in Bucks, it may be called the Risborough rock-bed. Fossils are generally abundant in it. The total thickness of this zone is from 100 to 120 ft.

*The Totternhoe Stone* has been described in several Survey Memoirs, and is a firm grey chalk which generally contains small scattered fragments of brown phosphate of lime, and frequently larger phosphatic nodules which are often greenish outside. In the area now under description it is only a thin band varying from 2 to 4 ft. thick, and it appears to die out westward in Berkshire, for it has not been traced beyond Chilton, which is a village just outside the western border of Sheet 254. The Totternhoe Stone has a gritty feel to the touch but is not a sandy chalk, neither is it calcified like the Risborough Rock, but its character is due to the quantity of even-sized fragments of shell which enter into its composition. *Rhynchonella mantelliana* (J. de C. Sow.) and *Kingena lima* (Defr.) are generally abundant in it.

[1] *Quart. Journ. Geol. Soc.*, vol. xxi, p. 398.

Above the Totternhoe Stone there is always a certain thickness of grey or greyish chalk, blocky and indistinctly bedded; this often passes up into whiter chalk, but the change is generally rapid, and sometimes the grey is divided from the whiter chalk by a bed of hard grey chalk containing phosphatic nodules and much disseminated phosphatic matter in minute grains. Of the whitish chalk there is generally about 50 or 60 ft., and the total thickness of the blocky chalk is from 80 to 100 ft. Fossils are not abundant in any part of it, but *Holaster subglobosus* (Leske), *Hol. trecensis* Leym., and *Metacanthoplites rotomagense* (Brongn.) may usually be found when there is a good exposure of the zone.

At the top of this zone is the band which has been called the "zone of *Belemnitella plena*," but it is of small thickness and its small fauna is not so different from that of the beds below as to be really of zonal importance, though it may be regarded as a sub-zone of the comprehensive zone of *Holaster subglobosus*. In the counties of Berkshire, Oxfordshire, and Buckinghamshire this sub-zone consists of two layers of grey shaly marl, separated by a bed of firm compact white chalk; each member varies in thickness from a few inches to 2 ft., but the whole band is seldom more than 3 ft. thick.

The lower marl band is the most variable of the three, being sometimes merely a seam or marly parting, but occasionally swelling to a foot or 18 in. The hard white course appears to be of constant occurrence in this district, and the upper marl is generally about a foot thick. This upper marl has yielded the characteristic Belemnite (*Actinocamax plenus* Blainv. sp.) more frequently than the other beds; but it will be convenient to speak of the whole band as "the Belemnite Marls." *Rhynchonella cuvieri* d'Orb. and a variety of *Rh. plicatilis* (J. Sow.) are not uncommon, and occur in all three layers, as also does a small globose variety of *Ostrea vesicularis* Lam.

## *Stratigraphical Details.*

### CHALK MARL.

### Zone of *Schloenbachia varians*.

Hard chalk marl can be seen at several points round the villages of Blewbury and Aston Upthorpe, but the only pit-section is near Sheencroft Farm, north-east of Blewbury. South of the farmstead there was in 1886 a small pit exposing about 7 ft. of soft grey marl, with included lumps of hard whitish glauconitic stone; this stone resembles some varieties of malmstone, is apparently very siliceous, the silica being probably organic, and the grains of glauconite are only just large enough to be seen without a lens. The marl in which the stony lumps occur passes down into a tough grey marl, and the tenant of the farm stated that it was much the same for another 8 ft. This exposure is not far from the outcrop of the Greensand (see p. 17).

North of Aston Tirrold the lower part of the Chalk Marl is concealed by valley gravel, and east of that village it is covered by alluvium, but the higher beds form the slopes of Lollington Hill south of the farm of the same name (spelt Loringdon on the old Ordnance Map); on the north-west side of this hill there is a small pit in tough blocky grey chalk, but this probably belongs to the zone of *Holaster subglobosus*, though the absence of fossils makes it uncertain.

Half-a-mile east of this hill and by the side of the road to Cholsey, is a larger pit which is certainly in the zone of *Schloenbachia varians*. It is about 12 ft. deep, in hard marly chalk with a course of softer marl about 4 ft. from the surface. Fossils are not abundant but *Schloenbachia varians* (J. Sow.), *Rhynchonella mantelliana* (J. de C. Sow.), and fragments of *Inoceramus* were found.

A very good exposure of the central part of the same zone is to be found in the railway cutting near Moulsford Station. This is about 20 ft. deep near the middle, the upper 10 or 11 ft. consisting of courses of marly chalk separated by layers of soft shaly marl ; below are about 8 ft. of soft grey marl with occasional lumps of harder material, resting on 2 or 3 ft. of very hard rocky chalk, weathering into lenticular lumps. Fossils are abundant throughout, especially *Schloenbachia varians* (J. Sow.), *Mantelliceras mantelli* (J. Sow.), and *Inoceramus latus* Goldf. and a *Rhynchonella* like *Rh. grasiana* d'Orb.

A large part of Cholsey stands on Chalk Marl, but eastward this passes beneath the gravel of the Thames valley. At the County Lunatic Asylum a boring reached the base of the Chalk Marl at 46 ft. from the surface.[1]

On the eastern side of the Thames the Chalk Marl emerges from beneath the gravel at Mongewell, and hard marly rock (Risborough rock) is visible at the spring heads at the upper end of the fine millpond. Thence the outcrop runs N.N.E. through Newnham Park and Crowmarsh, and the highest beds were exposed in the quarries on Crowmarsh Hill (see p. 25).

In Marsh Wood, N.E. of Crowmarsh, there are springs, but no good section, though the water probably issues from the rock. From the water-course a deep ditch has been dug northward, and at the north end of this, just one mile N.E. of Crowmarsh, I found a dry greyish marl full of green grains, so very glauconitic as to resemble the basement green marl ; but whether it is the basal marl brought up by a fault or a glauconitic layer higher up in the marl could not be determined.

In a field on the west side of Clay Lane, and about 6 furlongs S.W. of Ewelme Mill (Fyfield on the 1-in. map) a well was being dug in 1886 through soft clay marl into tough brownish sandy chalk or "clunch," with many fossils (see p. 27), the lowest beds then exposed, about 15 to 16 ft. from the surface, were hard and probably belonged to the top of the Risborough rock. This rock is exposed in a cutting by the roadside south of the mill house, and is here very compact, calcareous and hard. Below the millpond there is soft Chalk Marl in the watercourse.

At Ewelme we find the upper and firmer beds of the chalk which underlies the Totternhoe Stone exposed at the fine spring heads in the garden of the house above King's Pond. It is a greyish blocky chalk drying nearly white, and the same unfossiliferous chalk is seen in a small pit by the roadside north of Ewelme.

In Eyres Lane, north-west of Ewelme, soft marl is seen in the cutting overlain by a rubble of hard grey rock; if this indicates the outcrop of the Risborough rock it is at a level of 300 ft., whereas at Ewelme Mill its level is 245 ft., and we may assume the existence of a fault running along the hollow north of Ewelme with an upthrow of 50 ft. on the north-east side. The existence of such a fault is confirmed by following this contour northward. The next road cutting is overgrown, but the rock was found by picking into the bank. It supports the plateau of Scald Hill, its outcrop running to an old pit by the side of Rumbolds Lane, which is now overgrown but

[1] 'Water Supply of Berkshire' (*Mem. Geol. Surv.*), 1902, p. 57.

formerly exposed 5 or 6 ft. of very hard calcareous stone with *Schloenbachia varians* and other fossils (see p. 27).

Beyond and north of this old quarry the Chalk Marl is cut off by a fault which appears to run from Berwick Prior eastwards to Brightwell and Cuxham (see Fig. 2).

In 1899 Mr. Rhodes, visiting the area to collect fossils, found a new quarry open by the roadside near Rumbold's Copse, exposing about 10 ft. of grey chalk which probably belongs to the upper part of the Chalk Marl, and is not far below the Totternhoe Stone. He obtained a fair number of fossils from it, including *Schloenbachia varians* (J. Sow.), *Calycoceras naviculare* (Mant.), *Pecten (Æquipecten) beaveri* J. Sow., *Inoceramus latus* Goldf., *Rhynchonella grasiana* d'Orb., and a specimen of *Palæga carteri* H. Woodw.

At Brightwell Uppertown, above the spring-head, there is hard grey marly stone like weathered Risborough rock, with fragments of *Schloenbachia varians*, and in the bank by the cottages to the eastward there is a small exposure of shaly marl with hard concretionary lumps as large as a man's head, in which I found *Brachiolites sp.* (decomposed), *Inoceramus latus*, *Lima sp.* (? *echinata*), *Natica sp.* (? *gaultina*) and *Schloenbachia varians* (J. Sow.). Some other fossils were subsequently found here by Mr. Rhodes.

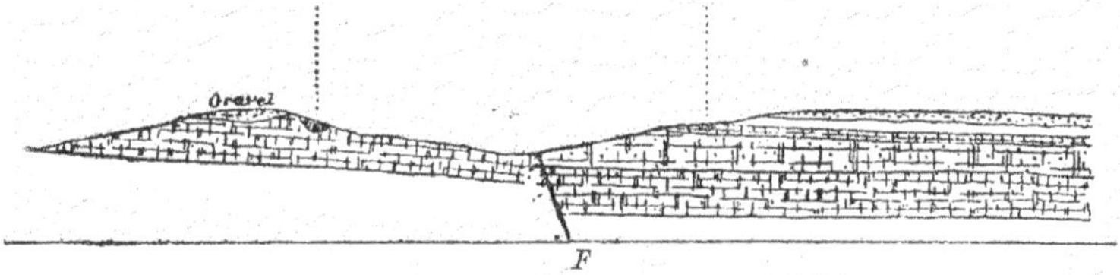


Fig. 2.—*Fault between Berrick Salome and Cuxham.*

*a* Malmstone. *b* Chalk Marl. *c* Risborough Rock. *d* Gravel. *F* Fault.

It would appear, therefore, that the Risborough rock continues to crop out along the 300 ft. contour which passes just above the spring-head. At Cuxham, south of Manor Farm, there are stronger springs, and here also I found fragments of the hard calcareous rock.

A little beyond this, near what is known as the middle mill, the outcrop appears to be cut by a fault, with a throw of some 40 ft., for the Watlington springs, which probably issue from the rock-bed, are above the 330 ft. contour. There is no good exposure at Watlington, but fragments of the rock are scattered over the field north-east of the church, and the upper beds of the rock are exposed at the bottom of the bank a quarter of a mile S.E. of Pyrton Church.

The wells in Pyrton are dug through soft greyish white marl, and find water at about 25 ft. down, but the basal glauconitic marl is nowhere exposed.

In Shirburn Park the outcrop of the Risborough rock is at the 360 ft. contour, and there are five springs at this level N.E. of the castle. The strike fault dies out with the north-easterly rise in the beds, and the base of the chalk marl is brought to the surface near the Model Farm north of the castle.

Soft grey marls are found round Moor Court, and by the roadside east of Nethercote, while the hard rock-bed can be seen above the fine springs which issue from beneath the main road at Lewknor. The feature made

Fig. 3.—*Section from Swyncombe Downs to Brightwell Baldwin.*

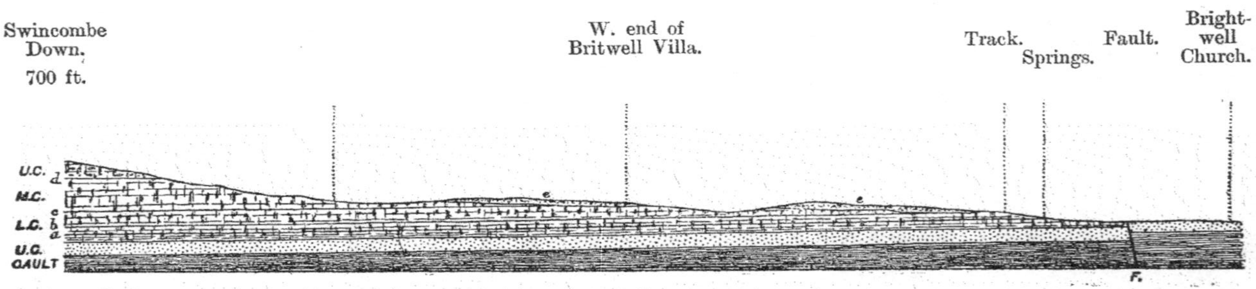


e. Plateau Gravel.
U.C. Upper Chalk } d Chalk Rock.
M.C. Middle Chalk } c Melbourn Rock.
L.C. Lower Chalk { b. Totternhoe Stone.
{ a. Risborough Rock.
U.G. Upper Greensand.
F. Fault.

**Distance about 3 miles. Vertical scale 1,200 feet to an inch.**

by the outcrop of this rock can be followed through the fields to the head of the fish-pond near Aston Rowant House.

The lowest beds of the Chalk Marl can be seen in the ditch by the road-side north of Aston Church, they consist of grey sandy marls passing down into hard, grey, gritty marl, with softer marl again just before the outcrop of the green sand is reached.

Returning to the outcrop of the Risborough rock, its course can be traced to the spring-head at Kingston Blount, and just above this, in the foundations for some outhouses, 5 or 6 ft. of hard marly stone were exposed in 1885; the same beds were also seen in a saw-pit on the north side of Crowell Church.

North of Crowell there are several strong springs, but the banks were not clear enough to expose the sub-soil, and the actual outcrop of the rock probably runs between them and the road.

At the point where the rivulets from these springs unite into one water-course there is soft Chalk Marl, but just below this there is dark green sand, and thence the base of the marl was traced with some difficulty to the north end of Chinnor.

The only exposure found at Chinnor was in the railway cutting at the station, where a few feet of rather hard white chalk were seen, and it was ascertained that the well was 45 ft. deep, being probably carried quite through the rock before finding water. The chalk seen here is not far below the Totternhoe Stone.

## Zone of *Holaster subglobosus*.

In the lanes south of Blewbury grey chalk with *Holaster subglobosus* (Leske) was found, but no trace of the Totternhoe Stone could be seen, neither was any good exposure of the Belemnite Marls seen when the district was mapped.

The central part of the zone, consisting of blocky grey chalk, is exposed in a pit near the main road, south of Aston Apthorpe. Similar chalk is seen in a pit on Lollington Hill to the eastward.

The outcrop of the Totternhoe Stone was found by the road to Cholsey, half a mile S.E. of Lollingdon Farm; it has the usual characters, contains many brown phosphatic nodules, and seems to be 3 or 4 ft. thick. Search was made for its outcrop in Halfpenny Lane, south-west of Moulsford Station, but without success, and the line drawn through Moulsford is entirely hypothetical.

South of Moulsford there appears to be a line of fault, for the hard yellow chalk of the *Rhynchonella cuvieri* zone occurs at a very much lower level on the south side of Moulsford Bottom than it does on the north side.

On the east side of the Thames the outcrop of the Totternhoe Stone is concealed for some distance by river-gravel, but the highest part of the zone can be seen near the "Leathern Bottle" Inn, north of Cleeve. There is a sort of cave here in which a spring rises, and the water is said to have been formerly saline, but the river water was let in either by accident or on purpose, and no saline taste can now be discerned; the chalk is in massive blocks and greyish white.

This part of the Lower Chalk occupies a considerable space to the east of South Stoke, round the slopes of White Hill and Watch Folly; there is a small pit more than half a mile east of Stoke and another on the north slope of White Hill, but no fossils were found in either pit, nor in the road-cutting south-west of Ipsden.

The summit of the zone is exposed in a small excavation by the roadside one and a-half miles east of South Stoke, the beds seen being:—

| | Ft. | In. |
|---|---|---|
| Melbourn Rock, hard and nodular - | 2 | 0 |
| Grey marl - - - - | 1 | 0 |
| Hard white rocky chalk - | touched | |

No outcrop could be found near Ipsden, but on White Hill, between

Ipsden and North Stoke, there is the following section in a pit by the roadside:—

| | | Ft. | in. |
|---|---|---|---|
| | Soil and chalk rubble - - - from 1 ft. to | 2 | 0 |
| Melbourn Rock | Hard nodular rocky chalk in regular beds, yellowish, but weathering nearly white - | 5 | 0 |
| Zone of *Act. plenus.* | Buff-coloured laminated marl with *Actinocamax plenus* (Blainv.) - 6 in. to | 1 | 0 |
| | Hard compact white chalk - 1 ft. to | 1 | 3 |
| | Greenish shaly marl, with *Ostrea vesicularis* Lam. and *Rhynchonella* - - about | 0 | 4 |
| Zone of *Hol. subglobosus.* | Firm blocky white chalk, with patches and mottlings of grey material near the top, giving it a brecciated appearance | 10 | 0 |

The strong spring at North Stoke cannot be far above the horizon of the Totternhoe Stone. At the south end of the wood by the side of the main road north of this spring-head is a pit exposing some 20 ft. of the firm white chalk belonging to the upper part of the zone, and beyond Mongewell Farm there is another pit in the same kind of chalk. At the bottom of the road-cutting north of this place there is rough-feeling grey chalk which cannot be far above the horizon of the Totternhoe Stone.

By the road along the south side of Newnham Park there is a small pit in greyish-white marly chalk containing *Metacanthoplites rotomagensis* (Brongn.) and *Inoceramus latus* Goldf. which can hardly be more than 30 ft. above the Totternhoe Stone, and the road cutting to the east of this gives a nearly continuous though shallow exposure of the greater part of the zone.

The large quarries by the side of the main road on Crowmarsh Hill are now abandoned, but in 1885 they exposed a nearly complete section of this zone, including the Totternhoe Stone and 18 ft. of the Chalk Marl below.

The highest face still shows about 30 ft. of nearly white blocky chalk, between the foot of this and the level on which the kiln stands there must be from 30 to 40 ft. of similar chalk, whitish above but passing down into grey, and containing *Metacanthoplites rotomagensis.* The bedding is obscure and irregular, but the general inclination is eastward.

In the two lower levels below the kiln the following section was seen in 1886:—

| | | Ft. |
|---|---|---|
| | *Middle Level.* | |
| Base of zone of *Holaster subglobosus.* | Rubble - - - | — |
| | Blocky grey chalk with oblique jointing - | 6 |
| | *Discoidea cylindrica* (Lam.), *Terebratula biplicata* J. Sow., *Teredo.* | |
| | Passing down into hard brownish-grey gritty chalk with many phosphatic nodules near the base (Totternhoe Stone) | 4 |
| Chalk marl. Zone of *Schloenbachia varians.* | Hard bluish-grey marl, breaking into blocks with a curved or semi-conchoidal fracture - | 8 |
| | *Lowest Level.* | |
| | Dull grey sandy stone, rather hard, but splitting into shaly layers and without phosphates | 2 |
| | Soft grey marl without definite bedding | 8 |
| | | 28 |

The dip appeared to be S.E. at about 2 deg. I was informed that the Totternhoe Stone was sometimes used for building barns and outhouses and for lining cisterns, etc.

North-east of the Crowmarsh Quarry and east of Coldharbour Farm the cutting for a cart road exposes the junction of the Lower and Middle Chalk, the succession being :—

| | | Ft. | In. |
|---|---|---|---|
| Brown gravelly soil | - | 1 | 0 |
| Rubble with pockets of gravel - | - | 2 | 0 |
| *Melbourn Rock*, hard nodular rocky chalk | | 3 | 0 |
| *Belemnite Marls.* | Soft grey laminated marl | 1 | 0 |
| | Hard compact white chalk - | 1 | 0 |
| | Thin seam of grey marl | 0 | 1 |
| Hard white chalk, seen for | - - - - | 5 | 0 |

Lower beds consisting of hard whitish blocky chalk are exposed in the limekiln quarry on Beggarsbush Hill, the depth of which is between 30 and 40 ft. This chalk contains very few fossils, Mr. Rhodes' search yielding only *Ostrea vesicularis* Lam., *Terebratula semiglobosa* J. Sow., a *Plocoscyphia*, and a fish bone.

The next exposure of this zone is in the cutting for Day's Lane, south of Ewelme, this is through the lowest beds and the Totternhoe Stone, but is much overgrown.

North-east of Ewelme part of the chalk surface is obscured by a deep covering of flint-gravel, and even where this is absent there are few exposures. The springs near Britwell Priory appear to rise from the grey chalk which overlies the Totternhoe Stone. At the spring-head south-east of Watlington, grey gritty stone was found which was taken to be Totternhoe Stone, though the exposure was too small for certainty. It is, however, on the line to which its outcrop would be shifted by the fault which is believed to run south-eastward from Cuxham. There is no traceable feature beyond this, but the line is drawn on by contour to the next exposure.

The white chalk, which forms the highest part of the zone of *Holaster subglobosus*, may be seen in the railway cutting south-west of Lewknor, and in a pit by the road south of that village. The same chalk is exposed in the cuttings near Aston Rowant Station.

By the side of the road less than half a mile south of Crowell there is an exposure of tough grey chalk, 3 or 4 ft. thick, resting on crumbling marl; this may be Totternhoe Stone or the bed seen in Crowmarsh Quarry 8 ft. below it. South-east of the point where this road crosses the Icknield way, the Belemnite Marls were seen dipping sharply into the hill side.

From the higher of two pits on the north-east side of the lane leading from Crowell to Crowell Hill, Messrs. C. P. Chatwin and T. H. Withers have lately obtained small subangular pieces of light brown, and of dark glauconitic, sandstone; also the following fossils :—

Exogyra haliotoidea (*J. Sow.*)
Ostrea vesicularis *Lam.*
Lima globosa (*J. de C. Sow.*)
Terebratulina striata *Dav.*
Discoidea cylindrica (*Lam.*)
Holaster trecensis *Leym.*
„ subglobosus (*Leske*)

## *Fossils from the Lower Chalk.*

The following is a list of the fossils which have been found at different localities in the area described. All have been obtained either by myself or the fossil collector, Mr. Rhodes, except those from Pyrton, which were found by Mr. Montague Smith, who presented them to the Woodwardian Museum, Cambridge, and those from Crowell above mentioned. The localities in the zone of *Schloenbachia varians* are indicated by letters thus: m[1] = Moulsford cutting, c = Crowmarsh, e = Clay Lane near Bensington, r = Rumbold Lane and Brightwell, p = Pyrton.

[1] A few (m) were found by Mr. Ll. Treacher and by Messrs. Chatwin and Withers.

| | *Zone of Schloenbachia varians.* | *Totternhoe Stone.* | *Zone of Hol. subglobosus.* | *Belemnite Marls.* |
|---|---|---|---|---|
| *PISCES.* | | | | |
| Beryx sp. - | — | - - | + | |
| *CRUSTACEA.* | | | | |
| Cytherella ? - - | r | | | |
| Palaega carteri *H. Woodw.* | r | | | |
| Pollicipes glaber *F. A. Roemer* - | e | | | |
| *CEPHALOPODA.* | | | | |
| Anisoceras armatum ? (*J. Sow.*) - | p | | | |
| Calycoceras naviculare (*Mant.*) | m r | | | |
| Hoplites falcatus (*Mant.*) - - | p | | | |
| Mantelliceras mantelli (*J. Sow.*) | m e p | — | + | |
| Metacanthoplites rotomagensis (*Brongn.*) | (m) | — | + | |
| Schloenbachia varians (*J. Sow.*) - | m e r p | | | |
| " coupei *Brongn.* | (m) e p | | | |
| Turrilites costatus *Lam.* - - | e r | | | |
| " mantelli *Sharpe* - - - - | r | | | |
| " tuberculatus *Bosc.* - | p | | | |
| Actinocamax plenus (*Blainv.*) - - | — | — | — | + |
| Nautilus elegans *J. Sow.* - - | e | | | |
| *GASTEROPODA.* | | | | |
| Aporrhais *cf.* carinata (*J. de C. Sow.*) | p | | | |
| Avellana cassis *d'Orb.* - - | p | | | |
| Natica vulgaris *Reuss* - - - | r p | | | |
| Pleurotomaria perspectiva ? (*Mant.*) - | (m) p | | | |
| " " [flat.] - | (m) | | | |
| Turbo sp. - - - - - | — | + | | |
| *LAMELLIBRANCHIATA.* | | | | |
| Aucellina gryphaeoides (*J. de C. Sow.*) - | p | | | |
| Cardita *cf.* tenuicosta (*J. de C. Sow.*) - | r | | | |
| Exogyra haliotoidea (*J. Sow.*) - | | | + | |
| Grammatodon carinatus (*J. Sow.*) [Cucullæa *Auctt.*]- - - | r p | | | |
| Hinnites sp. - - - - - - | p | | | |
| Inoceramus latus *Goldf.* (*non Mant.*) | m c e r p | | | |
| " striatus *J. de C. Sow.* | m | | | |
| " sp. - - | — | — | — | + |
| Lima aspera (*Mant.*) - - - | — | + | | |
| " (Mantellum) elongata (*J. de C. Sow.*) | e | | | |
| " " " *var.* echinata *Eth.* | r | | | |
| " (Plagiostoma) globosa (*J. de C. Sow.*) | p | + | + | |
| Ostrea vesicularis *Lam.* - - - - | c | — | + | + |
| Pecten (Æquipecten) beaveri *J. Sow.* - - - | r | — | + | |

| | Zone of *Schloenbachia varians.* | *Totternhoe Stone.* | Zone of *Hol. subglobosus.* | *Belemnite Marls.* |
|---|---|---|---|---|
| Pecten (Syncyclonema) orbicularis *J. Sow.* | m e r | | | |
| „ (Neithea) quinquecostatus *J. Sow.* - | c | | | |
| Pholadomya decussata *Phill.* - - | p | | | |
| Plicatula inflata *J. de C. Sow.* - | c r | — | + | |
| *BRACHIOPODA.* | | | | |
| Rhynchonella grasiana *d'Orb.* - - - | m c r | + | | |
| „ mantelliana (*J. de C. Sow.*) | m | + | | |
| „ plicatilis (*J. Sow.*) | — | — | — | + |
| „ schloenbachi *Dav.* | m | | | |
| Terebratula semiglobosa *J. Sow.* | m | + | + | |
| Terebratulina striata *Dav.* | r | | + | |
| *ECHINODERMATA.* | | | | |
| Discoidea subuculus (*Leske*) | p | | | |
| „ cylindrica (*Lam.*) | | | + | |
| Holaster subglobosus (*Leske*) - | — | — | + | |
| „ trecensis *Leym.* | | | + | |
| *SPONGIÆ.* | | | | |
| Pharetrospongia ? | r | | | |
| Plocoscyphia sp. - | m r | — | + | |

# CHAPTER VI.

## UPPER CRETACEOUS (*continued*).

### Middle Chalk.

By A. J. Jukes-Browne.

This division of the Chalk, as now defined by the Geological Survey, is that portion of the formation which lies between the Belemnite Marls and the Chalk Rock. It must be admitted that this is not a very good definition, because the Chalk Rock does not extend all over the English Cretaceous area; still an horizon which is probably equivalent to the base of that rock can be recognised with tolerable certainty throughout the south and south-east of England. In Berkshire, Oxfordshire and Buckinghamshire there is no difficulty, as the Chalk Rock is well developed, and the change from soft to hard nodular chalk is rapid, although there is no marked plane of division at the base of the rock. Moreover, it is certain that there is a change of fauna in the nodular passage beds at the base of the Chalk Rock.

The Middle Chalk is divisible with two zones, a lower one characterised by *Rhynchonella cuvieri* d'Orb. and *Inoceramus mytiloides* Mant., and an upper one characterised by *Terebratulina lata* Eth. and *Conulus subrotundus* Mant.

The zone of *Rhynchonella cuvieri* has the hard nodular chalk of the Melbourn Rock at its base, and this from its superior hardness makes a more or less well-marked feature which can easily be followed across the country. This is particularly the case on each side of the Thames valley, but in the north-eastern part of the area, the outcrop of the rock retires to the foot of the main escarpment and the feature to which it gives rise is less conspicuous. The Melbourn Rock is from 8 to 10 ft. thick, and it passes up into less nodular chalk, but beds with scattered nodules of harder chalk occur at intervals, and fragments of *Inoceramus* shell are common. In Oxfordshire this zone is probably about 60 ft. thick.

The zone of *Terebratulina* consists of firm homogeneous white chalk in thick beds, with occasional seams or layers of soft grey shaly marl. Such marly seams occur at variable distances from each other, sometimes there are two only a few feet apart and no other one for a thickness of 30 or 40 ft. *Conulus subrotundus* is generally to be found in the lower part of the zone, but is rare in the upper part. *Terebratulinæ* occur throughout, but in some localities they seem to be so rare that it is difficult to find a specimen.

Mr. Osborne White has examined many exposures of the beds at the junction of this zone with that of *Holaster planus*, not only in Oxfordshire and Buckinghamshire, but also in Berkshire and Hampshire, with the view of ascertaining whether the change of fauna takes place at or below the base of the Chalk Rock. He finds that a seam of soft marl always occurs at a distance of several feet below the base of that rock, and that the intervening chalk is always of a lumpy or nodular character.

North of the Thames valley this lumpy chalk is from 3 to 5 ft. thick, but when traced westward and southward from the Thames valley its thickness increases to 8 or 9 ft., always having what seems to be the same bed of marl at its base. In the chalk below this marl *Terebratulina lata* Etb. is common, and in the marl it is generally to be found, but above this horizon it is rare. In the lumpy chalk Mr. White has found *Terebratula carnea* J. Sow., *Ter. semiglobosa* J. Sow., *Spondylus spinosus* (J. Sow.), and *Micraster corbovis* Forbes. No specimen of *Holaster planus* or of *Echinocorys* has been found below the Chalk Rock; consequently we may infer that this lumpy chalk does belong to the *Terebratulina* zone, in spite of the rarity of that shell, and that the base of the Chalk Rock may be taken as the base of the overlying zone.

While this Memoir was passing through the press a paper by Messrs. Chatwin and Withers on "The Zones of the Chalk in the Thames Valley" was published.[1] They have found a few other species in the lumpy chalk, but their conclusions completely confirm the view above expressed.

In Berkshire the total thickness of the Middle Chalk is not more than 160 ft., but it seems to become rapidly thicker as it passes beneath the valley of the Thames, so that along the outcrop in Oxfordshire it is at least 200 ft., and probably more than that near Watlington.

It appears to dip very regularly beneath the Upper Chalk which crowns the main escarpment, and it crops out again inside the escarpment wherever the valleys have been cut down to a level which intersects the inclined plane of the Chalk Rock. Owing to an error which will be explained in the next chapter, these inlying tracts were not completely mapped when the area was surveyed, except in the north-east corner of the district; and owing to the same error it was supposed that the outcrop near Henley was due to a local periclinal flexure.[2] This is a mistake, and it is now known that the Henley outcrop is only a part of a large inlying tract, one arm of which stretches up the Assendon valley to and beyond Stonor, while the other arm passes up the Hambleden valley beyond Turville and Fingest to the heads of the two subsidiary valleys in the hills north of Ibstone (see the Map, Fig. 1, Frontispiece).

[1] *Proc. Geol. Assoc.*, vol. xx., p. 390 (1908).
[2] See 'Cretaceous Rocks of Britain,' vol. iii, p. 205.

## *Stratigraphical Details.*

### 1. Zone of *Rhynchonella cuvieri.*

The outcrop of the Melbourn Rock enters the area of this map about half a mile south of the main road through the village of Blewbury, and is exposed in a small excavation at a height of 323 ft. by the side of the road leading southward, and known as White Shoot. Still higher up the hill firm thick-bedded chalk is seen.

South of Aston this zone forms a bold escarpment which is furrowed by deep combes. The Melbourn Rock is exposed in a small quarry half a mile south of Aston Upthorpe.

Thence to the neighbourhood of Moulsford there are no exposures. On the north side of Kingstanding Hill the base of the zone runs at or near the contour of 300 ft., but on the south side hard chalk occurs at a lower level, and the outcrop of the Melbcurn Rock was not definitely ascertained, except to the south of Starveall Farm, whence a feature can be traced to the main road at Moulsford Bottom, which it crosses at about 240 ft. Thence a bold feature runs in nearly a straight line to the eastward, and seems to be produced by a fault which brings the hard nodular chalk of the *Rh. cuvieri* zone down against the softer Lower Chalk.

Along this line no exposure of the Belemnite Marls nor of the Melbourn Rock could be found, but on the main road south of Moulsford an old quarry on the 200 ft. contour exposes hard, yellowish, lumpy chalk, which seems to belong to the zone of *Rh. cuvieri* and is probably not far above the Melbourn Rock. No fossils were discovered in it, but the weathered face has a disintegrated rubbly appearance, owing to the weathering out of large irregular lumps of very hard chalk ; features which are also developed in the chalk of this zone at Cleeve, near Goring.

It was therefore inferred that the Melbourn Rock came to the surface on the slope between this road and the river, passing thence below the gravel, and reaching the level of the alluvium a little to the north of Streatley. A small pit in the field a quarter of a mile south-west of Streatley Farm exposed a few feet of similar hard chalk.

On the eastern side or the valley the lower part of the Middle Chalk is well exposed in the railway cuttings, and at several points along the river bank. The first cutting north of Goring Station is about 30 ft. deep ; the lowest beds in the centre are hard and full of fragments of Inoceramus, and form regular courses from 1 to 2 ft. thick ; the higher beds are less hard, and about half-way up the face a single egg-shaped flint was found, but the whole of the chalk has a yellowish tinge, and probably belongs to the zone of *Rh. cuvieri.*

An old pit by the bank of the river south of Cleeve Mill exposes the lower part of the zone ; hard, yellowish chalk, not regularly bedded, but consisting of irregular layers of loose, rubbly chalk, with lenticular masses of hard, shelly chalk, which stand out prominently from the weathered face of the pit. *Rhynchonella cuvieri* d'Orb. and *Inoceramus mytiloides* Mant. are common with some other fossils (see p. 54). Similar chalk but in more massive beds, very hard and resembling the top of the Melbourn Rock, occur at the level of the water nearer the mill, and the same hard beds are visible in an excavation in the railway cutting about 150 yards north of the bridge.

At the north end of the Temple Grounds, where one would expect to find still lower beds, there is a quarry with the following section :—

| | Ft. |
|---|---|
| Soil and rubble - - - - - - | 2–4 |
| Soft white chalk, with finger-like flints - - - - - | 8 |
| Hard, rocky, yellowish chalk, with *Inoceramus mytiloides* Mant. | 10–14 |

The dip is northerly, very slight, but sufficient to show that Cleeve stands on a low east and west anticline.

The actual outcrop of the Melbourn Rock cannot be seen, but it evidently forms the base of the ridge which runs from Spring Farm to Grove Farm. By the roadside N.W. of Grove Farm fragments of hard, rocky chalk are frequent, and thence the feature can be traced to Mile End Hill, three-quarters of a mile N.E. of Grove Farm, where a small excavation shows the base of the rock (see p. 24).

Though carefully searched for by Braziers Park and Ipsden, no traces of the rock could be found, nor any exposure of the Cuvieri Zone. The line, however, was run by feature to the quarry on White Hill, near North Stoke, where the base of the rock and about five feet of the hard nodular rock tself are again exposed, at a height of about 270 feet.

The plateau north of this spot which presents such a fine steep scarp to the valley of the Thames has evidently withstood detrition, partly in consequence of the hardness of this chalk, and partly, also, perhaps, because there may be an east and west syncline beneath it.

No exposures occur till we reach Coldharbour Farm, near Crowmarsh; here a deep cutting has been made, exposing the base of the rock (for section see p. 26).

The outcrop is thence concealed by gravel, and though it comes to the surface again in the valleys east of Ewelme, no sections were seen. By Brockholes Covert, near Britwell House, there is a pit in hard white chalk containing *Rhynchonella cuvieri* d'Orb. and *Conulus subrotundus* Mant., and there is another small pit east of the house.

From Britwell the zone can be traced by feature to the quarry by the main road east of Watlington Farm, which exposes about 24 ft. of firm, yellowish-white chalk, with layers of loose nodular chalk; the lowest beds are very hard, the higher contain *Rhynchonella cuvieri* and *Inocerami* in abundance.

Three-quarters of a mile farther to the N.E., and near the farm called Beechwood, there is a small pit exposing the top beds of the Melbourn Rock, and about the same distance north of this is a much larger quarry, in which an interesting section is exposed; the knoll is caused by a syncline or pericline, and the beds rise from the centre of the quarry both to the N.W. and S.E. In the centre the succession is:—

| | Ft. |
|---|---|
| Hard nodular chalk, without distinct bedding, and weathering into irregular lumps - - - - - - - - | 14 |
| Hard compact rock, not nodular, breaking into large blocks - | 4 |
| Hard rocky chalk with a central nodular or concretionary band, but the whole forming one bed - - - - - - - - | $1\frac{1}{2}$ |
| Compact chalk, in thin beds, with marly layers, the chalk being brittle and smooth - - - - - - - - - - | 3 |
| Band of grey marl containing nodules or lumps of white chalk of all sizes, and some of them coated with green matter, thickness varying up to - - - - - - - - - - - - | $0\frac{1}{2}$ |
| Hard compact chalk in thick massive beds - - - - - - | 4 |
| | 27 |

These lowest beds are undoubtedly the top part of the Melbourn Rock, but the marl above it is an unusual feature, as is also the intercalation of smooth compact beds, with the nodular chalk of this zone.

The rock is next seen in the road cutting south of Lewknor, but the section is overgrown; thence its outcrop strikes E.N.E. below Beacon Hill, and it can be seen below the hedge on the London road close to Aston Rowant Station; here it is at a level of 475 ft., and it continues to rise till it reaches the 500 ft. contour.

It is cut through in the road cutting half a mile S.E. of Crowell, the zone of *Actinocamax plenus* dipping sharply into the hill, and overlain by very hard white rock, weathering into rounded nodular lumps, and passing up into hard yellowish nodular chalk with *Inoceramus mytiloides* Mant.

## 2. Zone of *Terebratulina.*

Near Blewbury the chalk of this zone forms Riddle Hill and Upthorpe Down, but there is no good exposure of it. It passes eastward to King-tanding Hill, where the higher part of it, about 50 ft. below the Chalk Rock, is exposed by the road on the south side of the wood, and consists of soft white chalk, with *Terebratulina lata* Eth. and *Discoidea dixoni* Forbes.

This chalk forms the slopes of the deep combes or dry valleys which open eastward into the valley of the Thames, but the only other exposure is in a pit by the roadside north of Streatley; this shows some depth of firm white chalk, containing fragments of *Inoceramus* and a few scattered flints in the upper part. Here Dr. Barrois noted "*Ostrea vesicularis*, Lk., *Spondylus speciosus* Sow., *Inoceramus* near to *labiatus*."[1]

On the east side of the Thames Valley the upper part of the Middle Chalk forms a tract of some width from near Cleeve to Harcourt Hill, north of Nuffield, but there are very few exposures of it.

A pit near the barn south of Grove Farm shows part of the *Terebratulina* zone, but yielded no fossils. There is a small pit in soft white chalk half a mile east of Ipsden, and another by Poors Farm, north of Hailey.

The uppermost beds are exposed in Garson's Lane, one mile E.S.E. of Ipsden, and consist of tough white chalk in solid beds without flints, and containing a band of soft chalk, in which a species of *Inoceramus* is abundant. The same beds occur in the road-cutting S.E. of Well Place Farm.

The next exposure is a large quarry near Potter's Farm, south-east of Ewelme, the lower level of this shows about 15 ft. of soft white blocky chalk without distinct bedding, and containing a few large flints scattered through its mass; about 12 ft. above the floor there is an irregular layer of grey laminated marly chalk; the upper level is about 20 ft. deep, but disused and obscured. The Chalk Rock is just exposed at the upper end of the working. *Terebratulina lata* Eth. is very abundant in the lower beds, and *Rhynchonella plicatilis* (J. Sow.), *Spondylus spinosus* (J. Sow.) and remains of *Inoceramus* were obtained higher up. The chalk has been burnt for lime.

An old pit by the roadside, over half a mile east of Britwell House, shows firm white chalk with *Terebratula semiglobosa* J. Sow. and *Terebratulina lata* Eth.

The uppermost beds of the Middle Chalk are exposed in small pits at the foot of the hill south of Howe Farm. At the top of one of these excavations, on the western side of the road, and a few feet below the Chalk Rock, there are irregular marly seams, beneath which *Terebratulina lata* is common. *Plicatula barroisi* Peron, *Inoceramus spp.*, and *Ostrea* also were noticed. The lowest beds are seen in the upper level of the quarry by the Henley road, S.E. of Watlington; they consist of firm white chalk with *Conulus subrotundus* Mant., etc., but I did not notice any flints. From 25 to 30 ft. are exposed.

The higher beds are seen again in a pit in the combe on the north side of Watlington Hill, by the roadside below Bald Hill, near Lewknor, and on the main London road up Aston Hill.

They are quarried by the side of the road north-east of Kingston Grove, the pit showing firm white chalk with curvilinear planes of division, but no definite bedding. No flints nor any fossils were seen.

The same kind of chalk, but at a lower horizon, is dug in a small pit just above the zone of *Rhynchonella cuvieri* south-east of Crowell, and again in a larger pit on Chinnor Hill, near Carlton House, but no fossils were found at either locality.

Mr. H. J. Osborne White contributes the following notes on exposures in the valleys north of Henley:—

"The highest beds of the *Terebratulina* Zone are exposed at the bottom of the well-known Chalk Rock quarry by the lower lodge at Henley Park. The chalk is white, harsh to the touch, and rather massive, though it assumes a flaggy character on weathering. A description of the whole section is given below (p. 42).

"At the eastern end of a deep (and, for the most part, overgrown) quarry a quarter of a mile north of the Telegraph Office at Greenlands there is a clear section, 25 to 30 ft. deep, of beds rather lower in the zone. The

1 'Recherches terr. Crétacé Sup. de l'Angleterre,' etc. (1876), p. 147.
Messrs. Chatwin and Withers have lately published a detailed description of this section. *Proc. Geol. Assoc.* vol xx (1908), p 405.

chalk is rather soft, white, blocky, and almost flintless, and contains a grey marly band, a few inches wide, about 20 ft. above the floor of the working. Fossils seem exceedingly scarce, and an examination of well-weathered surfaces revealed nothing of interest.

"The highest beds are seen again at the kiln, one furlong north of Westfield Farm, west of Medmenham. The lower part of the large pit there shows about 15 ft., and formerly showed 30 ft. or more, of firm to soft, widely-jointed, white chalk, containing marly seams, and becoming greyish and tougher towards the top. There are a few thin lenticles of flint with ill-defined boundaries. A well-marked marl seam, 1 in. to 3 ins. thick (with another less distinct just below it), separates these beds from the harder lumpy chalk which merges into the Chalk Rock, 5 ft. to 6 ft. above. *Terebratulina lata* occurs sparingly at the top, but becomes common in the whiter beds about 10 ft. down. The other forms collected by me are:

Scalpellum fossula? *Darw.*
Inoceramus cuvieri *J. de C. Sow.*
,, sp. (near to I. mytiloides *Mant.*)
Ostrea vesicularis *Lam.*
Dimyodon nilssoni (*Hag.*) [Plicatula sigillina *S. P. Woodw.*]
Spondylus spinosus (*J Sow.*)
Terebratula semiglobosa *J. Sow.*
,, carnea *J. Sow.*
Terebratulina striata *Dav.*
Micropora hippocrepis *Goldf.*
Cyphosoma radiatum *Sorig.*
Discoidea dixoni *Forbes*
Micraster corbovis *Forbes*
Porosphæra globularis (*Phill.*)

"*Terebratulina*-chalk from this quarry was used, with brick and flints, in the construction of the church and some of the older houses in the village of Medmenham, and has been worked for building on the Danesfield estate in recent years. The beds which were being quarried for building-stone in the lowest part of the pit, below ground-water level, in 1889-90, are of compact, widely-jointed white chalk, with fine interlacing venules of a pale-grey tint.

"Although it forms the bottom and lower side-slopes of the Assendon Valley, from Henley up to Turville Park, the Middle Chalk has been little worked in that combe. The best sections are those in a series of small excavations, above and below the 400 ft. contour-line, in the banks of the pretty hollow lane which runs from Hollandridge Farm south-south-eastward to the main road between Pishill and Stonor. These show white and greyish-white blocky chalk, with occasional seams of marl, and grey bands of closely-set, impersistent, marly laminæ. Fragments of an *Inoceramus* like *I. mytiloides* are abundant in places, but other fossils are not easily found. *Prionotropis woolgari* (Mant.), [casts], *Inoceramus brongniarti* J. de C. Sow., *I. cuvieri* J. de C. Sow., *Ostrea sp.*, *Spondylus sp.*, and *Terebratulina lata* Eth. have been got from the lower beds, which are probably 50 to 60 ft. below the top of the zone.

"A tongue of Middle Chalk runs up the western fork of the Assendon Valley, from Middle Assendon nearly to the Rifle Range at Maidensgrove Scrubs. There is a little opening in greyish, flaggy chalk by the roadside at Bix Bottom Farm, and smaller exposures occur in the road-banks farther up the valley.

'Between Mill End, at its mouth, and Wormsley, north-west of Ibstone, at its head, the broader valley of Hambleden is cut from 50 ft. to upwards of 100 ft. into the *Terebratulina* Zone. It is possible that the *Rhynchonella cuvieri* beds immediately underlie the Valley Gravel near Turville and Fingest.

"Road-bank exposures of the higher zone are numerous, and fair sections can be seen to the south-west of Flint Hall; on the end of the Turville Hill spur, west of Fingest; in the hollow lane running south-eastward from North End to Turville; and at the corner of the copse near the barns a quarter of a mile south-east of Ibstone Manor Farm. None of these sections has been closely examined, and the only fossils I can record are :—

Prionotropis woolgari (*Mant.*)
Inoceramus sp. (near to I. mytiloides *Mant.*)
Inoceramus spp.
Ostrea sp.
Plicatula barroisi *Peron*
Terebratula semiglobosa *J. Sow.*
Terebratulina gracilis *var.* lata *Eth.*
Eschara lamarcki *Hag.*
Proboscina ?
Asteroidea
Hemiaster minimus (*Ag.*)
Ventriculites mammillaris *T. Smith.*"

# CHAPTER VII.

## UPPER CRETACEOUS (*continued*).

### UPPER CHALK.

BY A. J. JUKES-BROWNE AND H. J. OSBORNE WHITE.

Only a portion of the Upper Chalk enters into the structure of the area under description, the zones comprised in this portion being (1) the zone of *Holaster planus*; (2) that of *Micraster cortestudinarium*; (3) the greater part of the zone of *M. coranguinum*. The maximum thickness coming in between the base of the Chalk Rock and the surface on which the Eocene outliers rest being from 250 to 260 ft. The beds which compose the three zones above mentioned will be treated separately in the following pages.

### 1. Zone of *Holaster planus*.

The zone of *Holaster planus*, which includes the beds known as the Chalk Rock, is now taken as the lowest part of the Upper Chalk. The line engraved on the map is that of the base of the Chalk Rock, and for all practical purposes it represents the base of the zone. The base of the Chalk Rock was so regarded in the general Memoir on the Upper Chalk (1904), though it was admitted that this rock passed down into the underlying chalk.

Before the publication of that Memoir, Dr. A. W. Rowe had shown that the base of the zone could be fixed in all the south coast sections by the incoming of *Micraster præcursor* Rowe and *M. leskei* (Desm.), but as none of these sections include a typical development of Chalk Rock, they did not settle the question of the zonal position of that rock in those counties where it is found.

As stated on p. 30, further examination of the lumpy chalk which underlies the Chalk Rock in this area has only confirmed the view that the base of the rock is the base of the *H. planus* zone, for in spite of the careful search made by Mr. Osborne White in every exposure, no specimen of *H. planus*, *Micraster leskei*, or *Echinocorys scutatus* Leske has rewarded his efforts, neither have any been found by Messrs. Chatwin and Withers. The Chalk Rock forms a band 3 or 4 ft. thick of hard cream-coloured limestone, containing grains of glauconite and including one or more layers of hard green-coated nodules, which consist of compact yellowish and slightly phosphatic chalk. It generally contains some of the characteristic fossils; its upper surface is usually a layer of the nodules.

Above the rock-bed is a variable thickness (12 ft. to 22 ft.) of nodular chalk, consisting of hard lumps embedded in a matrix of soft mealy chalk. Solid flint nodules occur at intervals, and generally along planes parallel to the bedding, but are set wide apart; frequently, also, there are some small scattered globular flints with thick white rinds. In these nodular beds *Micraster præcursor* is usually so abundant that I have referred to them in my note books as the "Micraster Beds."

At the top of these beds there is in some places a thin bed of hard yellowish limestone, resembling the Chalk Rock below; this bed is from 12 to 18 in. thick, and occurs at or near the top of the zone. At other localities, however, no such rock-bed can be found, and the beds at the top of the zone exhibit much variety, changing within short distances. Mr. Osborne White has recently visited most of these exposures, and has been able in most cases to fix the approximate junction of the zone with that of *Micraster cortestudinarium* by the evidence of the varietal changes presented by the tests of *Micraster præcursor*, and he finds the total thickness of the zone of *Holaster planus* to be about 26 ft.

It has also been found that a serious omission was made in the mapping of the district, but the fact was not discovered until the colour-printed map (Sheet 254) was ready for issue. It would appear that the late Mr. J. H. Blake failed to obtain evidence of the existence of the Chalk Rock along the valleys which open out north of Assendon and Hambleden. He had mapped it as passing under the floor of these valleys near the places above mentioned, but Mr. O. White has found that the level of the Chalk Rock rises northward and allows a broad strip of Middle Chalk to be exposed around Stonor in the one valley and around Turville and Fingest in the other, as already stated (see p. 30).

Since this omission could not be corrected on the colour-printed issue of the Geological Survey Map, it has been decided to show the correction on a map specially prepared for this Memoir by Mr. Osborne White, whose exploration of the country led him to discover the mistake. If this map is compared with the published 1 in. sheet, it will be seen that the height to which the outcrop of the Chalk Rock rises north of Fingest and Ibstone approximates to the level at which it was found by me to occur near Radnage; this mapping proves that the base of the Upper Chalk forms an inclined plane which is practically uninterrupted by flexures, so that where valleys have been cut down below the level of this plain inlying tracts of Middle Chalk have been exposed.

The outcrop of the Chalk Rock between Henley and Hambleden is not, therefore, due to any anticlinal or periclinal flexure, as the original mapping led one to suppose, but is merely part of the exposure of the inclined plane above mentioned.

A. J. J.-B.

### *Stratigraphical Details.*

## 1. Exposures Along the Escarpment.

### By A. J. Jukes-Browne.

The outcrop of the Chalk Rock, and consequently that of the zone of *Holaster planus*, enters the area of Sheet 254 about three-quarters of a mile south of Blewbury, and it runs out along the promontories of Cholsey Downs, Moulsford Down, and Thurle Down. In this district surface exposures are occasionally visible, but no quarry or clear road-section was seen during the progress of the Survey in 1885.

After passing southward by Streatley and Basildon,[1] it re-enters our area on the eastern side of the Thames Valley to the south-east of Cleeve, and in the road-cutting about a quarter of a mile east of Fairfield House, close to the 300 ft. contour line, the following section was seen in 1885:—

| | | Ft. |
|---|---|---|
| | 7. Soft white chalk with a thin layer of flint at base | 5 |
| Zone of *Hol. planus.* | 6. Hard yellowish limestone in loose lumps | 1 |
| | 5. Micraster Beds, consisting of hard lumps of white chalk embedded in soft powdery chalk, with a few scattered flints - - - | 12 |
| | 4. Chalk Rock. Hard white limestone without green grains passing down into very hard compact yellowish rock with much glauconite, and containing several layers of green-coated nodules | 5 |
| | 3. Hard rocky chalk passing down in rough nodular chalk - - - | 2 |
| | 2. Layer of soft shaly marl - - - - - - | $0\frac{1}{4}$ |
| | 1. Firm bedded greyish-white chalk, nodular in places | 4 |
| | About - | 30 |

The Micraster Beds are here less productive than usual, but I found *Echinocorys scutatus* Leske, *Spondylus spinosus* (J. Sow.), and *Terebratula carnea* J. Sow. Messrs. Chatwin and Withers have obtained some fossils from the Chalk Rock here.[2]

The outcrop of the rock can be traced round the hill into the combe by Elvendon Farm, where I found fragments of it in the bank by the roadside.

It is exposed in Red Lane three-quarters of a mile N.W. of Woodcote at a level of about 380 ft., showing in the bank as hard, cream-coloured rock, much of it with green grains from 3 to 4 ft. thick.

In Garson's Lane, one mile E.S.E. of Ipsden, the base of the rock is seen in the pits by the roadside at a level of 426 ft., a hard, yellow limestone with green nodules, and 50 yds. further east there is white, powdery chalk, with hard lumps (the Micraster Beds).

No sign of the rock could be found in the road-cutting above Well Place, though fragments of the rock had been used for making the road, and there is a small grass-grown pit at the level where its outcrop should be. Neither could it be found east of Hailey, but on Batchelor's Hill, east of Wicks Wood, its outcrop was found at a level of about 460 ft.

In 1886 it was not exposed along any of the roads between Warren Hill and Ewelme Downs, and hardly a fragment of it could be found; thus showing how completely the outcrop of the rock may be hidden beneath soil and wash. Mr. Osborne White, however, states that a little of this rock, with its green-coated nodules, can be seen between the top of the talus and the turf at the upper end of the pit in *Terebratulina*-chalk to

1 See 'Geology of the Country Around Reading' (*Mem. Geol. Surv.*), p. 9.
2 See *Proc. Geol. Assoc.*, vol. xx, p. 393.

the south-east of Potter's Farm. He also points out that the outcrop of this bed must cross the main road from Henley to Oxford nearly half a mile further to the north-west, and some 80 ft. lower than is shown on the 1-in. map; for a small roadside excavation, rather more than a furlong north-west of Ambrose Farm, shows yellowish nodular chalk which clearly belongs to the upper half of *Holaster planus* zone. In view of these facts it is probable that the Upper Chalk boundary line has been mapped at too high a level throughout that part of the escarpment which lies between Warren Hill and Ewelme Downs.

From the pit near Ambrose Farm, just mentioned, Mr. White obtained:—

Inoceramus cuvieri *J. de C. Sow.*
" sp.
Dimyodon nilssoni (*Hag.*) [= Plicatula sigillina *S. P. Woodw.*]
Lima (Plagiostoma) hoperi (*Mant.*)
Spondylus spinosus (*J. Sow.*)
Rhynchonella plicatilis (*J. Sow.*)
Terebratula carnea *J. Sow.*
" semiglobosa *J. Sow.*
Cardiaster cotteauanus *d'Orb.*
Cidaris sp.
Echinocorys scutatus *Leske* (gibbous form)
Holaster placenta *Ag.*
" planus (*Mant.*)
Micraster præcursor *Rowe*
Pharetrospongia strahani *Sollas.*
Porosphæra globularis (*Phill.*)
Ventriculites infundibuliformis *S. Woodw.*

The examples of *Micraster præcursor* have "slightly inflated" inter-poriferous areas.

On White Hill, north-west of Swincombe, the rock crops out in the road at a level of 521 ft. Passing out of the combe it juts out westward in the fine promontory of Swincombe Down, and fragments may be found near the end of the Danish entrenchment, which runs along the north side of this hill.

In Howe Lane (south of Howe Farm) a little of the firm, lumpy, beds at the base of the rock was visible (in 1906) in a ditch near the bottom of the hill, at about 550 ft., and a few feet higher than the pit in *Terebratulina*-chalk noticed in the last chapter (p. 33).

On Watlington Hill fragments of the rock occur in the deep ditch for carrying off rain-water just above the 600 ft. contour, and there are old pits by the roadside, whence I was told that "rag" for road mending and for building walls had been dug. The lower part of the wall in the yard of the Fox and Hounds Inn, Watlington, is built of Chalk Rock, which was also used for road material when the parish had the care of the main road to Henley.

In the roadway on the north side of Shirburn Hill the outcrop is not clear, but what appeared to be the upper bed of the rock, as seen further north, was observed at 700 ft. overlain by chalk with a layer of tabular flint.

Near the road to Lewknor, down Bald Hill, and only just above the 600 ft. contour line, there is a pit exposing Chalk Rock, but though seemingly undisturbed, this can hardly be *in situ*, and is probably part of a mass that has slipped down the steep slope.

On the east side of Beacon Hill, which is the next promontory, fragments of rock were found at a level of about 670 ft., and this is exactly the level of its outcrop at the quarry by the London road on Aston Hill.

At this quarry there was (in 1885) an excellent section showing two separate beds of Chalk Rock, with the Micraster Beds between them, as follows:—

| | | Ft. |
|---|---|---|
| | 4. Soft, thin-bedded chalk, with a layer of flints in the middle - - - - - - - - - | 6 |
| Upper Rock. | 3. Course of compact, yellowish rock, with green nodules - - - - - - - - - | 1 |
| Micraster Beds. | 2. Dull, white, lumpy chalk, consisting of soft, mealy chalk, with hard lumps and many fossils, especially near the base, where it contains much glauconite and a layer of peculiar flints - - - - - - | 14 |
| Chalk Rock. | 1. Loose, rubbly rock, with green grains, green-coated nodules, and many fossils. Passing down into - | 1 |
| | Compact cream-coloured rock, with another layer of nodules about 2 ft. down (base not seen) - - | 4 |
| | | 26 |

The bedding is nearly horizontal, and small slips or faults are numerous. There is a remarkable abundance of glauconite grains, the base of the chalk above the lower rock being in some places more of a greensand than a chalk, and they extend upwards for at least 2 ft. The flints at this horizon are rough and carious with hollow spaces.

After 1885 this quarry remained for many years unworked and obscured by talus, but was re-opened in 1907, and was visited by Messrs. Chatwin and Withers, who carefully collected from each horizon and have kindly placed the results at our disposal.

In the chalk above the upper rock they found :—

Micraster præcursor *Rowe* (of the M. cortestudinarium zone type)
Holaster placenta *Ag.*
Rhynchonella limbata (*Schloth.*)
Terebratula carnea *J. Sow.*
" semiglobosa *J. Sow.*
Echinocorys scutatus *Leske* (bits)
Inoceramus sp.
Serpula fluctuata *J. Sow.*
Membranipora elliptica *Reuss*
Ventriculites impressus *T. Smith*

From the upper rock (3) they obtained the following :—

Micraster cortestudinarium (*Goldf.*)
" præcursor *Rowe*
Holaster sp. (bits)
Bourgueticrinus sp.
Pleurotomaria perspectiva (*Mant.*)
Spondylus spinosus (*J. Sow.*)
Inoceramus sp.
Dimyodon nilssoni (*Hag.*) [= Plicatula sigillina *Woodw.*]
Terebratula carnea *J. Sow.*
Serpula granulata *J. de C. Sow.*
Plocoscyphia convoluta (*T. Smith*)

From the Micraster Beds they obtained :—

Micraster præcursor *Rowe*
" leskei (*Desmoul.*)
Holaster planus (*Mant.*)
" placenta *Ag.*
Echinocorys scutatus *Leske*
Spondylus spinosus (*J. Sow.*).
" latus (*J. Sow.*)
Terebratula carnea *J. Sow.*
" semiglobosa *J. Sow.*
Nautilus sp.
Oxyrhina sp.
Serpula fluctuata *J. de C. Sow.*
" plana *S. Woodw.*
Stomatopora gracilis (*M. Edw.*)
Proboscina augustata (*d'Orb.*)
Berenicea disciformis (*Hag.*)
" *cf.* phlyctænosa *Reuss*
" *cf.* contracta *Seeley*

From the Chalk Rock (No. 1) the following were obtained :—

Nautilus sp.
Crioceras ellipticum (*Mant.*)
Barbatia *cf.* geinitzi (*Reuss*)
Inoceramus sp.
Pachydiscus peramplus (*Mant.*)
" sp.
Avellana *cf.* humboldti *Müller*
Cerithium cuckhamsliense *Woods*
Dentalium turoniense *Woods*
Solariella gemmata (*J. de C. Sow.*)
Trochus beroscirensis *Woods*
Turbo geinitzi *Woods*
Pleurotomaria sp.
Septifer lineatus (J. de C. Sow.)
Spondylus spinosus (*J. Sow.*)
Terebratula carnea *J. Sow.*
" semiglobosa *J. Sow*
Rhynchonella cuvieri *d'Orb.*
" plicatilis (*J. Sow.*)
Serpula ampullacea *J. de C. Sow.*
Holaster planus (*Mant.*)
Echinocorys scutatus *Leske*
Parasmilia centralis (*Mant.*)
Cliona cretacea (*Portlock*)

It may be mentioned that five specimens of the *Nautilus* were obtained from the topmost loose glauconitic bed.

About a furlong east of the above, and on the lower road from Warren Farm, there is an old limekiln in a slipped mass of chalk where a similar section must have formerly been visible, but it is now much hidden by talus.

The upper beds are seen again in a small pit by the road east of Kingston Grove; here the section is :—

| Zone | Bed | Ft. |
|---|---|---|
| Zone of *Micraster cortestudinarium.* | 5. White chalk with scattered flints, with a line of large flint nodules at the bottom - | |
| | 4. Thin-bedded white chalk with two thin seams of tabular flint - - - | 4½ |
| Zone of *Holaster planus.* | 3. Hard cream-coloured rock with green nodules and ferruginous concretions; well defined above, but merging below into the underlying chalk - - - - - - | 1–2 |
| | 2. *Micraster Beds.*—Hard yellowish chalk which becomes nodular below, consisting of hard lumps, embedded in soft greyish mealy chalk - - - - - - - | 14 |
| | 1. The *Chalk Rock* is hidden by talus, but is seen on the bank on the other side of the road. The upper rock bed (3) was still well exposed in 1906, but, though closely examined, yielded no fossils of interest. | |

The base of the rock is exposed at the top of an old pit on the cart road down Crowell Hill; and another good section of the whole group occurs in a pit by the main road on Chinnor Hill, exactly on the contour line of 700 ft. Fig. 4 is a sketch of this pit.

*Fig. 4.—View of a Quarry on Chinnor Hill.*

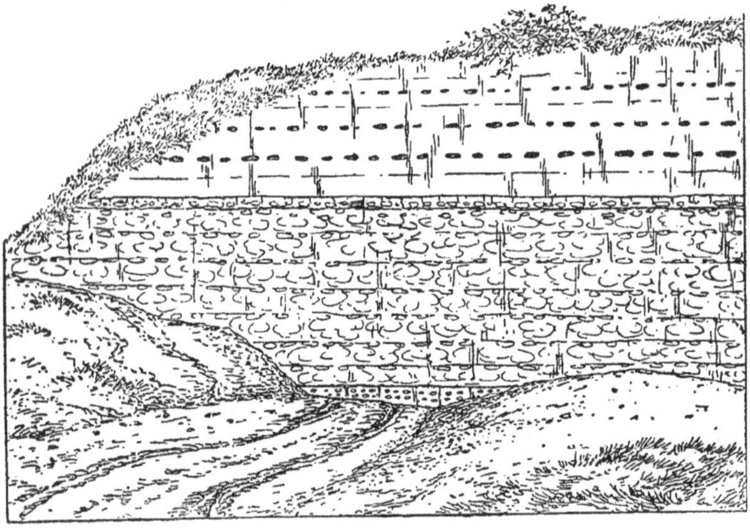

| | | Ft. |
|---|---|---|
| 4. White chalk with three layers of flints, the lowest nearly continuous - - - - - - - - - - - - | | 10 |
| Zone of *Holaster planus.* | 3. Hard yellow rock, upper surface well defined, merging below into underlying chalk - | 1 |
| | 2. Dull white mealy chalk with lumps of hard white chalk (Micraster Beds) - - - | 14 |
| | 1. *Chalk Rock.*—Hard yellowish limestone with a layer of green-coated nodules at the top and many green and black grains, seen for - - - - - - - | 2 |
| | | 27 |

The contrast between the dull white tint and lumpy character of the Micraster Beds, and the whiteness, even fracture, and bedded character of the chalk above the upper rock bed is here very marked.—(A. J. J.-B.)

## 2. Exposures on the Dip-slope.

### By H. J. O. White.

The best sections in the inliers are those in the quarries opened on the northern slope of the Thames Valley, between Henley and Medmenham. Some account of their lithological features has been given in previous publications of the Geological Survey and elsewhere, but, so far as I am aware, no attempt to determine the limits of the *Holaster planus* Zone in them has hitherto been made.

The results of my own recent examination of these sections are not wholly satisfactory, for distinctive fossils are very scarce near the bottom of the zone, while the highest beds are in no case sufficiently accessible to enable one exactly to locate the junction with the overlying *M. cortestudinarium* chalk. As Mr. Jukes-Browne has stated in the introduction to this chapter, the upper limit of this zone is unmarked by any persistent lithological feature.

In a disused quarry at the lower lodge of Henley Park, and about a mile north of the town, the subjoined section is seen :—

| | | Ft. |
|---|---|---|
| Zones of *Micraster cortestudinarium* and *Holaster planus* (junction probably between 1 and 3 ft. from base). | 6. White chalk, with many flints, in somewhat irregular courses; a well-marked flinty band, iron-stained in places, at the base. *Micraster præcursor* Rowe - - - | 12 |

| | | Ft. |
|---|---|---|
| Zone of *Holaster planus.* 25½ ft. | 5. Irregularly blocky to lumpy and nodular white to greyish-white chalk, consisting of hard, ill-defined lumps in a firm to soft matrix. Some solid flint-nodules, disposed in more or less widely-open courses, and becoming notably scarcer downwards; also scattered, small, globular flints, with thick, friable, white rinds. A more definite layer of flints at the base. *Inoceramus cuvieri* J. de C. Sow.; *I. spp.*; *Ostrea vesicularis* Lam.; *O. spp.*; *Plicatula barroisi* Peron; *Dimyodon nilssoni* (Hag.); *Spondylus latus* (J. Sow.); *S. spinosus* (J. Sow.); *Rhynchonella plicatilis* (J. Sow.); *R. reedensis* Eth.; *Terebratula carnea* J. Sow.; *T. semiglobosa* J. Sow.; *Stomatopora gracilis* Edw.; *Serpula spp.*; *Cyphosoma radiatum* Sorig.; *Echinocorys scutatus* Leske (gibbous and ovate forms); *Holaster placenta* Ag.; *Micraster cortestudinarium* (Goldf.); *M. præcursor* Rowe; *Ventriculites mammillaris* T. Smith; *V. alcyonoides* Mant. - | 21½ |
| | 4. *Chalk Rock*—Hard, yellowish, crystalline limestone, with grains of glauconite and layers of green-coated nodules, one of the latter occurring at the top. Upper limit clearly marked, but not the lower. Crooked, branching borings abundant. *Heteroceras sp.*; *Scaphites geinitzi* d'Orb.; *Cerithium sp.*; *Pleurotomaria perspectiva* (Mant.); *Cuspidaria caudata* Nilss.; *Inoceramus sp.*; *Spondylus spinosus* (J. Sow.); *Teredo amphisbœna* (Goldf.); *Trapezium trapezoidale* (F. A. Roemer); *Rhynchonella plicatilis* (J. Sow.); *Terebratula carnea* J. Sow.; *Echinocorys scutatus* Leske; *Serpula granulata* J. de C. Sow.; *Micraster sp.* (casts of a small form); *Parasmilia centralis* (Mant.); *Plocoscyphia convoluta* T. Smith - - - - - - - | 4 |
| Zone of *Terebratulina* (top). | 3. Hard to firm, lumpy, greyish-white chalk, with a few small, angular, green concretions; becomes irregularly blocky towards base. *Inoceramus cuvieri* J. de C. Sow.; *I. sp.*; *Ostrea vesicularis* Lam.; *Terebratula semiglobosa* J. Sow.; *Terebratulina striata* Dav.; *Discoidea dixoni* Forbes [=*D. minima*] *Micraster corbovis* (Forbes) - | 4½ |
| | 2. Seam of grey marl, 1 to 3 ins. *Inoceramus sp.* (fragments). | |
| | 1. White to greyish-white chalk; firm, harsh to the touch, and widely jointed (flaggy where weathered). Contains a thin seam of grey marl 1 ft. below top. *Inoceramus sp.*; *Ostrea sp.*; *Spondylus spinosus* (J. Sow.); *Terebratulina lata* Eth.; *Cidaris serrifera* Forbes; *Porosphæra globularis* (Phill.) - - seen | 9 |

The examples of *Micraster præcursor* found at many horizons between the base of bed (5) and the top of the flint band at the base of bed (6) are all of *H. planus* Zone types, and a comparison of their features with those of the tests of the same species obtained from beds at the junction of the *H. planus* and *M. cortestudinarium* Zones elsewhere leads me to think that the upper limit of the former zone in this section occurs between 22 and 24 ft. above the upper surface of the Chalk Rock. Tests with distinctly "sub-divided" areas—presumably from the middle or upper part of bed (6)—are found on the talus. I have met with no "sutured" forms at a greater height than 3 ft. above the Rock. It may be mentioned that the example of *Stomatopora gracilis* recorded from bed (5) is attached to a cast of Micraster, at the upper surface of the Chalk Rock.

Chalk Rock, about 3 ft. thick, is exposed at the top of the steep bush-grown talus in the old quarry north of Greenlands Telegraph Office, but I have seen no fossils in it.

At the quarry north of Westfield Farm, to the west of Medmenham, where chalk was worked for building-stone in 1899, and is now burnt for lime, the following succession is exhibited :—

| | | Ft. |
|---|---|---|
| Zones of *Micraster cortestudinarium* and *Holaster planus* (junction probably 5 or 6 ft. above base of bed 8). | 9. White chalk with many bands of rather small flints. A conspicuous course of flints at the base | 10 ? |
| | 8. White chalk with fewer flints; a well-marked course of flints at the base. *Echinocorys scutatus* Leske (gibbous form) *Micraster præcursor* Rowe | 12 ? |
| Zone of *Holaster planus*. | 7. White, irregularly-blocky to lumpy chalk (with an ill-defined nodular layer 8 in. from the top) passing down into a greyish and more nodular chalk with some thin marly partings. Flints—far apart in courses and scattered—are not uncommon near the top, but scarce towards the bottom. Small globular flints with friable rinds occur throughout. *Inoceramus cuvieri* J. de C. Sow.; *I. sp.*; *Ostrea vesicularis* Lam.; *O. sp.*; *Spondylus latus* (J. Sow.); *S. spinosus* (J. Sow.); *Rhynchonella plicatilis* (J. Sow.); *Terebratula carnea* J. Sow.; *T. semiglobosa* J. Sow.; *Terebratulina striata* Dav.; *Serpula sp.*; Bryozoa; *Echinocorys scutatus* Leske (gibbous, ovate, and other forms); *Holaster placenta* Ag.; *H. planus* (Mant.); *Micraster cortestudinarium* (Goldf.); *M. præcursor* Rowe; *Ventriculites sp.* | 17½ |
| | 6. *Chalk Rock:* Hard, cream-coloured crystalline limestone, with two well-marked and other less distinct layers of green-coated nodules. The junction with bed below is marked in some places by a joint-plane. *Baculites bohemicus* Fritsch and Schloen.; *Scaphites geinitzi* d'Orb.; *Cerithium saundersi* Woods; *Pleurotomaria perspectiva* (Mant.); *Avellana sp.*; *Trochus berocscirensis* Woods; *T. schlüteri* Woods; *Turbo sp.*; *Arca* cf. *galliennei* d'Orb.; *Spondylus latus* (J. Sow.); *Inoceramus sp.*; *Rhynchonella plicatilis* (J. Sow.); *R. reedensis* Eth.; *Stomatopora gracilis* (M. Edw.); *Terebratula carnea* J. Sow.; *T. semiglobosa* J. Sow.; *Echinocorys scutatus* Leske; *Micraster præcursor* Rowe (small) Echinoids, thin-tested,—seen only in section. *Parasmilia centralis* (Mant.); *Ventriculites sp.* | 4 |

| | | Ft. |
|---|---|---|
| Zone of *Terebratulina* (top). | 5. Hard, lumpy chalk, passing down into firm, massive chalk with fine, undulate, marly laminæ - - | $5\frac{1}{2}$ |
| | 4. Seam of grey marl; 1 to 3 in. | |
| | 3. Greyish-white, firm chalk, with a thin seam of grey marl at the base - - | 1 |
| | 2. Greyish, firm, massive chalk, becoming whiter downwards. Flints in thin lenticles and horned and finger-shaped nodules | 13 |
| | 1. Firm, white chalk - - seen for | 3 |

The fossils found in beds (1-5) have already been mentioned (see above, p. 34).

The beds have a slight dip to the south-east.

The rate of intraformational change in the test-features of *Micraster præcursor* (and *M. cortestudinarium*) appears to be the same as in the Henley Park section. The highest test of *Micraster præcursor* obtained by me occurred about 19 ft. above the Chalk Rock. It is of the "moderately inflated" type. I think it unlikely that the upper limit of the *Holaster planus* Zone here is situated less than 22 ft. above the top of the rock.

In his description of the *M. cortestudinarium* Zone sections of Oxfordshire and Buckinghamshire, Mr. Jukes-Browne (Cret. Rocks of Britain, vol. iii, p. 216) quotes the following note by Mr. W. Hill:—"'A quarry at Medmenham, north of the church, shows some 40 to 45 ft. of chalk, but the face is inaccessible except near the base. The lower 10 ft. consisted of lumpy chalk of irregular texture, and above this is a layer of large flints rather widely separated. The next 20 ft. contained many scattered flints and one prominent flint floor. The highest part consists of massive chalk with large rough flints in irregular layers.'" From the lumpy beds below the first well-marked flint course, 10 to 12 ft. above the floor of this pit, I have obtained:—*Inoceramus cuvieri* J. de C. Sow.; *I. sp.*; *Terebratula semiglobosa* J. Sow.; *Rhynchonella plicatilis* (J. Sow.); *Serpula ilium* Goldf.; *S. plana* S. P. Woodw.; *Bourgueticrinus* (columnal); *Cidaris sceptrifera* Mant.; *Echinocorys scutatus* Leske (rather small gibbous and ovate forms); *Metopaster parkinsoni* (Forbes); *Micraster præcursor* Rowe ("slightly inflated" type, 5 and 12 ft. above floor of pit) *Porosphæra globularis* (Phill.) and *Plocoscyphia convoluta* (T. Smith).

These lowest beds are, therefore, in the zone of *H. planus*, and it is probable that the flint course at their top is a continuation of that seen at the base of bed (8) in the section at the kiln, described above. Fragments of Chalk Rock were thrown out of a shallow well by the entrance to the quarry, and other pieces (probably nearly *in situ*) were formerly to be seen at the spring beside the Marlow road, a little to the south-east.

On the south side of the Thames, débris of the Chalk Rock can be seen at many points in the banks of the Henley-Remenham-Aston lane, but there is no clear exposure of the limestone in position. There are remains of a shallow working on the outcrop of the rock north of Remenham Lodge.

A good section of the upper part of the zone is shown in the pit east of Remenham Church. Since Mr. Jukes-Browne's Memoir on the Upper Chalk appeared (in 1904), this pit has been reworked for chalking the soil on the gravel-terrace just above, and the character of the beds can be better made out.

| | | Ft. |
|---|---|---|
| | Gravel. | |
| Zones of *Micraster cortestudinarium* and *Holaster planus* (junction is probably within 5 ft. of base of this bed). | 2. White chalk, rather soft at top, but becoming lumpy downwards, and containing ill-defined bands of harder, yellowish, lumpy and nodular chalk. Solid flints occur sparingly in open courses. The small globular flints, noticed in the preceding sections, are abundant, and there are some very thin seams of tabular flint.<br>A conspicuous floor of tabular flint at the base.<br>The following are from lower half of bed:—*Pleurotomaria perspectiva* (Mant.), *Inoceramus cuvieri* J. de C. Sow., *I. sp.*, *Spondylus spinosus* (J. Sow.), *Cidaris sceptrifera* Mant., *Echinocorys scutatus* Leske (gibbous, ovate, and other forms), *Holaster placenta* Ag., *Micraster cortestudinarium* Goldf., *M. præcursor* Rowe, *Plocoscyphia convoluta* (T. Smith), *Ventriculites infundibuliformis* S. P. Woodw., *V. mammillaris* T. Smith- - about | 18 |
| Zone of *Holaster planus*. | 1. Mixed blocky and lumpy white chalk, with undulate grey marly seams, passing down into a rough chalk, consisting of firm to hard nodules in a soft marly matrix. A few flints, scattered and in open courses.<br>Fossils: the same species as in bed (2), with the following additions:—*Corax sp.*, *Ostrea hippopodium* Nilss., *Terebratula carnea* J. Sow., *T. semiglobosa* J. Sow., *Proboscina crassa* (Röm.) var. *elevata* Greg., *Serpula cincta* Goldf., *S. plana* S. P. Woodw., *Bourgueticrinus*, *Pentacrinus*, *Parasmilia centralis* (Mant.), *Coscinopora quincuncialis* (T. Smith), *Porosphæra globularis* (Phill.) exposed | 11 |

*Micraster præcursor* is very abundant in bed (1), especially in a band 3 ft. from the top. The tests are mostly of the "feebly" to "moderately inflated" types; "sutured" forms being—so far as I have seen—limited to the lower half of the bed. In bed (2) I have found only three examples of the species, all within 2 ft. of the tabular flint-seam at the bottom. Two of these are of upper *H. planus* Zone types, while the third approaches the "sub-divided" form of the *M. cortestudinarium* Zone. Mr. H. L. Hawkins has one very broad test of *M. cortestudinarium* from a higher horizon.

I infer that the junction of the two zones lies somewhere between 2 and 5 ft. above the base of bed (2). *Holaster placenta* is very common in both divisions of the section. With the exception of one example of *Micraster cortestudinarium*, some tests of *Holaster placenta* and *Echinocorys scutatus*, a cast of *Pleurotomaria perspectiva*, and some sponge-impressions, all the fossils above recorded were obtained from chalk which I refer to the zone of *Holaster planus*.

In the Assendon Valley there are two good sections of chalk belonging to this zone. One of these is in the quarry at the back of the cottages rather more than a furlong south-east of Middle Assendon Saw Mill. The greater (and more fossiliferous) part of the chalk here is in the zone of *M. cortestudinarium*, but, as the beds about the upper limit of the *H. planus* Zone are easily accessible, it will be well to describe the section on this page. It shows;—

| Zone | Bed | Ft. |
|---|---|---|
| Zone of *Micraster cortestudinarium.* | 3. White blocky chalk, with many courses of solid and spongeous flints (mostly the former), some layers of tabular flint, and many sub-parallel oblique veins of the same. The lowest persistent layer of tabular flint in the section is taken as the base of this division about | 25 |
| | 2. Irregularly-jointed white chalk, with scattered harder lumps of yellow tint (2 to 3 ft.), passing down into lumpy and nodular yellowish chalk ($\frac{1}{2}$ to $1\frac{1}{2}$ ft). There is an irregular course of solid flints $1\frac{1}{2}$ ft. from the top, and another, less distinct, at the base. Fossils from this division, and from the lowest third of that above, include:— *Lamna appendiculata* (Ag.), *Ptychodus mammillaris* Ag., *Inoceramus brongniarti* J. de C. Sow., *I. cuvieri* J. de C. Sow., *Ostrea hippopodium* Nilss., *O. sp.*, *Spondylus spinosus* (J. Sow.), *Kingena lima* (Defr.) (rare), *Terebratula carnea* J. Sow., *T. semiglobosa* J. Sow., *Terebratulina striata* (Dav.), *Rhynchonella plicatilis* (J. Sow.), *Onychocella depressa?* Hag., *Proboscina radiolitorum* (d'Orb.), *Reptoceritites rowei* Greg., *Serpula ilium* Goldf., *S. gordialis* var. *serpentinus* Schloth., *Bourgueticrinus*, Asteroidea, *Pentacrinus*, *Cidaris sceptrifera* Mant., *Echinocorys scutatus* Leske (gibbous form), *Holaster placenta* Ag., *H. planus* (Mant.), *Micraster cortestudinarium* (Goldf.), *M. præcursor* Rowe, *Plocoscyphia convoluta* (T. Smith), *Ventriculites mammillaris* T. Smith - | $3\frac{1}{2}$ |
| Zone of *Holaster planus.* | 1. Lumpy and nodular yellow chalk, with harder bands and undulate marly partings. One of the hard bands, about 5 ft. from the top, is rich in remains of *Holaster placenta* and sponge casts. Nodular flints, mostly small and far apart in courses, and scattered. A few impersistent seams of tabular flint in the upper part. *Inoceramus cuvieri*, *I. spp.*, *Ostrea* sp., *Spondylus spinosus*, *Rhynchonella plicatilis*, *Terebratula carnea*, *T. semiglobosa*, Asteroidea, *Cidaris sceptrifera*, *Holaster placenta*, *Micraster præcursor*, *Coscinopora quincuncialis*, *Plocoscyphia convoluta*, *Porosphæra globularis* (Phill.), *Ventriculites radiatus* Mant. seen | $17\frac{1}{2}$ |

The Micrasters found up to within 18 in. of the flints at the base of bed (2) are of the "feebly" to "moderately inflated," *H. planus* Zone types. Those obtained at and above that horizon are all of the "sub-divided" and "strongly inflated" forms characteristic of the zone of *M. cortestudinarium.*

The beds have an apparent dip of 5 or 6 deg. to the north-west, and are broken by small faults. The Chalk Rock has been worked at this spot[1] but it is no longer visible.

In the western bank of Hollandridge Lane, and just within the adjoining field, at a spot rather less than three furlongs north-east of Pishill Church, there are some small and shallow excavations in the Chalk Rock, and in the lumpy, glauconitic beds immediately below it.

In the beds below the rock I noticed only *Terebratula carnea* J. Sow. and chips of thin *Inoceramus*-shell. From the rock itself, and chiefly from

1 W. Whitaker, 'Geology of the London Basin' (1872), p. 48.

the topmost layer of green nodules, which is directly overlain by the thin rubbly soil, Mr. Treacher and I have obtained :—

*Crioceras ellipticum* (Mant.), *Solariella gemmata* (J. de C. Sow), *Turbo geinitzi* Woods, *Spondylus latus* (J. Sow.), *Rhynchonella plicatilis* (J. Sow.), *R. reedensis* Eth., *Terebratula semiglobosa* J. Sow., *Reptoceritites rowei* Greg., *Micraster sp.* (casts of small form, common,? *M. leskei*), *Parasmilia centralis ?* (Mant.).

Casts and moulds of gasteropods are fairly abundant, but most of them are too much decayed by soil-water to admit of identification.

The Chalk Rock here forms a distinct terrace on the end of the Holland-ridge spur, the ground rising rather sharply across the edges of the very flinty *M. cortestudinarium* beds to the north.

A small opening in the eastern bank of the lane in the lower part of the slope above this terrace shows greyish, lumpy chalk, with bits of thin *Inoceramus* and of *Echinocorys*, and probably belonging to the upper part of the *H. planus* Zone.

Similar chalk, with *Terebratula carnea*, has been worked just above the outcrop of the Chalk Rock by the roadside at Pishill Bank. The excavation is very small, and the chalk in a dirty and rubbly condition.

In the Hambleden Valley I know of only one clear section of the *H. planus* beds, though the Chalk Rock may be seen in the banks and gutters of most of the roads which cross its outcrop, and projects through the talus in a degraded pit to the east of Calstrope, north of Hambleden.

By the side of the road leading up from the hamlet of Pheasants Hill to Rockwell End, and close to the former place, there are two small pits. The higher and more north-eastern pit is in the *M. cortestudinarium* Zone, and will be described under the appropriate heading. The lower pit shows :—

| | Ft. |
|---|---|
| 2. Greyish-white, lumpy and nodular chalk (in a semi-rubbly condition), containing a few flints of the usual *H. planus* Zone types. *Inoceramus cuvieri* J. de C. Sow., *I. sp.*, *Spondylus spinosus* (J. Sow.), *Terebratula carnea* J. Sow., *T. semiglobosa* J. Sow., *Micraster corbovis* Forbes ("sutured"), *M. præcursor* Rowe ("sutured"), *Coscinopora quincuncialis* (T. Smith), *Ventriculites mammillaris* T. Smith - - - - - - - - - | 8 |
| 1. *Chalk Rock.* Hard, cream-coloured, massive limestone, with grains of glauconite and green-coated nodules. *Heteroceras sp.*, *Solariella gemmata* (J. de C. Sow.), *Rhynchonella plicatilis* (J. Sow.), *Terebratula carnea*, *Parasmilia centralis* (Mant.) - - - - - - - exposed | 2 |

Very little of the rock is visible. In 1905 it was being worked for road-metal.

H. J. O. W.

## The Zone of *Micraster cortestudinarium.*

### By A. J. Jukes-Browne and H. J. Osborne White.

This zone is not a very satisfactory sub-division from a stratigraphical point of view. In this part of England it does not present any distinctive lithological features, while its fauna is scanty, and does not seem to include any species that are absolutely restricted within the limits of the zone. Practically, it can only be separated from the zones above and below by the varietal mutations of the Micrasters—*i.e.*, *M. præcursor* and *M. cortestudinarium.* Mr. Osborne White has succeeded in collecting a sufficient number of specimens to enable him to recognise the lower limit of the zone where it is exposed; but he has not been able to ascertain its exact upper boundary at any locality, for exposures at this horizon are few and fossils are rare.

The lower beds of this zone are exposed in most of the quarries where the Chalk Rock is seen, but exposures of the higher part are less frequent.

The lower beds, which may be about 20 ft. thick, consist for the most part of compact firm white chalk; but there are sometimes beds of lumpy, and even of rough, nodular chalk, which weather to a yellowish tint on the surface. Flints occur in layers of large solid nodules, black inside, with thin rinds.

The higher beds, which may be 30 ft. thick, consist of soft white blocky chalk, in which flints are very numerous, occurring both in layers and scattered through the mass. The majority of them are spongeous and cavernous, and small globose flints (often hollow) are very common; continuous seams and layers are also frequent. A. J. J.-B.

Along the escarpment of the Chilterns the lowest beds of this zone are poorly exposed in the upper part of the degraded Chalk Rock quarry at Chinnor Hill (p. 41), where the following fossils were collected by Messrs. Chatwin and Withers:—*Inoceramus sp.*, *Ostrea vesicularis* Lam., *Pecten cretosus* Defr., *Serpula fluctuata* J. de C. Sow., *S. ilium* Goldf., *Micraster præcursor* Rowe, *Ventriculites sp.* Higher beds are exposed in a road-side pit E. of Manor Farm, on Chinnor Hill; the chalk contains regular courses of solid and spongeous flint-nodules, and some layers and veins of tabular flint. Here Messrs. Withers and Chatwin obtained *Dimyodon nilssoni* (Hag.), *Spondylus latus* (J. Sow.), *Idmonea alipes* Greg., *Proboscina radiolitorum* (d'Orb.), *Holaster placenta* Ag., *Micraster cortestudinarium* (Goldf.), *Echinocorys scutatus* Leske, *Serpula ilium* Goldf., *S. fluctuata* J. de C. Sow., *S. plana* S. Woodw.

The single example of *M. cortestudinarium* found is of a type characteristic of beds about the junction of the *M. cortestudinarium* and *M. coranguinum* Zones. The horizon at which it occurred is estimated by Messrs. Withers and Chatwin to be about 20 ft. above the top of the *Hol. planus* Zone.

The best sections occur in the neighbourhood of Henley, and some of these have already been described (*see* pp. 42 to 47).

To the south of Bix Rectory, west of Middle Assendon, there is a little opening in white chalk with many spongeous flints, and a prominent layer of carious tabular flint. Here I have noticed :—*Inoceramus cuvieri* J. de C. Sow., *Ostrea* sp., *Rhynchonella reedensis* Eth., Asteroid ossicles, *Cidaris clavigera* König, *Echinocorys scutatus* Leske (a tall, gibbous form, and portions of a thin test), *Pentacrinus*, and *Pharetrospongia strahani*. This pit is probably 50 to 60 ft. higher than the outcrop of the Chalk Rock about one furlong distant to the north-east.

In a wood by the lane leading from Mill End to Rotten Row, and rather less than half a mile from the former place, a quarry shows about 50 ft. of white chalk with many courses of carious and solid flints, and some layers of tabular flint near the bottom. Small globular flints abound between 10 and 15 ft. above the floor of the pit at the north-western end. Some disturbance, which appears to be of a strictly local order, has given the beds the rather high dip of 10° to the south-west. After *Inoceramus cuvieri*, the commonest fossil in the lower beds is *Cidaris clavigera.* Examples of *Micraster præcursor* Rowe, of forms characteristic of the *M. cortestudinarium* zone, were found at 7½ and 16 ft. above the lowest flint-course seen in the section. The other fossils found by me in the lowest third of the section are:—*Inoceramus involutus* J. de C. Sow., *Lima hoperi* (J. Sow.), *Ostrea sp.*, *Spondylus latus* (J. Sow.), Bryozoa, *Serpula plana* S. P. Woodw., *S. cincta* Goldf., *S. pentangulata* S. Woodw., *Cidaris serrifera* Forbes, *Echinocorys scutatus* Leske (gibbous form, large), *Metopaster parkinsoni* (Forbes), *Plinthosella squamosa* Zitt., *Porosphæra globularis* (Phill.).

The highest beds, which are out of reach, are probably in the zone of *Micraster coranguinum.*

A small but fossiliferous section of the lowest beds of the *M. cortestudinarium* Zone is seen in the higher of the two pits near Pheasants Hill, before mentioned (p. 48). This is about 13 ft. deep, and shows irregularly-jointed, greyish-white chalk, with a course of rather small solid flints near the top and bottom, and some scattered flints between. Six to seven feet above the floor of the pit there is a band of harder, yellowish, lumpy and nodular chalk, with impressions of sponges.

The fossils found are:—*Inoceramus cuvieri* J. de C. Sow.; *Inoc. sp.*, *Ostrea hippopodium*? Nilss.; *Spondylus spinosus* J. Sow.; *Terebratula carnea* J. Sow.; *Thecidium wetherelli*? Morris; *Berenicea sp.*, *Stomatopora gracilis* (M. Edw.); *Serpula plana* S. P. Woodw.; *Cidaris serrifera* Forbes; *Echinocorys* (bits), *Micraster cortestudinarium* (Goldf.); *M. præcursor* Rowe; *Plinthosella squamosa* Zitt.; *Plocoscyphia convoluta* (T. Smith); *Ventriculites mammillaris* T. Smith.

The Micrasters found occurred between 4 and 8 ft. above the floor of the pit, which is probably rather more than 20 ft. higher than the top of the Chalk Rock exposed in the other excavation, a short distance to the south-west.

Three furlongs east of Rose Hill, and about seven furlongs south-by-east of Medmenham Abbey, there is a field-pit, 18 ft. deep, in white chalk abounding in spongeous and tabular flints. The beds dip about 5° to the south-east. A little hard, yellow, lumpy chalk is seen near the floor of the working.

The following forms have been found here:—*Inoceramus cuvieri*, *Inoc. sp.*, *Spondylus latus* (J. Sow.); *Rhynchonella reedensis* Eth.; *Terebratula semiglobosa* J. Sow.; *Terebratulina striata* Dav.; *Berenicea sp.*, *Cribrilina sp.*, *Entalophora virgula* (Hag.); *Eschara sp.*, *Homœosolen aculeatus* d'Orb.; *Membranipora navicularis* Reuss; *Membranipora sp.*, *Micropora hippocrepis* Goldf.; *Nodelea durobrivensis* Greg.; *Onychocella sp.*, *Proboscina fasciculata* (Reuss); *Reptoceritites rowei* Greg.; *Reptomultisparsa sp.*, *Siphoniotyphlus tenuis* (Hag.); Asteroid ossicles, *Bourgueticrinus*, *Cidaris hirudo* Sorig.; *Echinocorys scutatus* Leske (gibbous form); *Helicodiadema fragile* (Wilt.); *Micraster* (bits); *Coscinopora sp.*; *Doryderma sp.* H. J. O. W.

### 3. Zone of *Micraster coranguinum.*

The chalk of this zone contrasts with that of the zones below in being of a much more uniform character. It is white and blocky, of a soft texture, fine and homogeneous in grain, but high up in the zone one or more thin and rather hard lumpy beds occur. There are also occasional thin seams of grey marl, which are probably persistent over a large area.

The flints in this zone are generally solid and often large, occurring in frequent and regular courses; there are, however, some beds with carious and hollow flints and some impersistent floors of agglutinated nodules, while a few solid continuous layers ("tabular" flints) occur in the lower part of the zone.

Mr. Osborne White has not found it possible to determine the exact junction of this and the underlying zone, for the exposures at points where it might be expected to be visible are few and small, and the chalk about this horizon is poor in fossils, so that it has not been possible to obtain a sufficient body of evidence.

Assuming that the thickness of the zone of *M. cortestudinarium* is about 50 ft. thick, the maximum thickness of the zone of *M. coranguinum* will be 190 ft., but the higher beds are gradually cut out by the transgressive overstep of the Eocene deposits.

In the small tract of Upper Chalk which comes into the south-western corner of the area (Berkshire) no exposures were seen of this part of the Chalk.

Along the frontal slope of the Chiltern Hills only two quarries were seen in the beds above mentioned.

The first of these was nearly two miles east of Ipsden Church, and S.E. of Homer Farm; here is a small quarry in firm white chalk with two layers of large flint nodules, but no fossils could be found.

On Watlington Hill, and about a quarter of a mile south-west of Portways, is a pit in rather hard whitish chalk with a band of hard yellowish rocky chalk; flints are very abundant both in layers and as scattered nodules, most of them contain cavities with remains of sponges, some are pink, others grey or nearly black. This chalk must be at a height of about 750 ft., and, therefore, about 140 ft. above the base of the Chalk Rock. A. J. J.-B.

Mr. H. L. Hawkins, of Reading, has found *Micraster coranguinum*, spines of *Cidaris*, and a few other common fossils of this zone in a pit near Coates Farm (west of Howe Hill), and in small workings by the road side near the top of Gangsdown Hill, north of Nuffield.

On the dip slope there are many sections and exposures of beds belonging to this zone. I have examined a selection of these (numbered 1 to 11), and have indicated the fossils found in each of them in the *M. coranguinum* zone column in the general list at the end of this chapter.

1. A small pit close to the boundary of the Reading Beds, a little east of the church at Turville Heath, shows about 12 ft. of white chalk with rather small nodular flints of grey and pale mauve tints. A band of firm to rather hard, yellowish, lumpy chalk occurs near the base of the section, and a thin seam of grey marl a little higher up. *Kingena lima* (Defr.) and *Bourgueticrinus* are common. One very thin test of an ovate *Echinocorys scutatus* Leske was found.

The beds are traversed by a fault with a throw of 18 in. to the south-west, and by pipes filled with brown stony clay, and with mottled clay and sand.

2. There are a few poor exposures, about a foot deep, of soft to very hard white and greyish-white chalk, containing thick-rinded grey and violet flints, round the edges of old pits amongst the furze to the east of Russell's Water. The beds here are probably at or about the same horizon as those in pit (1).

3. The southern of two big pits in the east side of the combe (Harpsden Valley) west of Rotherfield Peppard, shows about 20 ft. of fine white chalk, containing many courses of large, solid flints, and some hollow, "mealy" flints of spherical form, about the size of a cricket-ball. Spines and plates of *Cidaris hirudo* Sorig. and *C. clavigera* König are common.

The northern pit seems to be at about the same horizon, but is less fossiliferous.

The chalk in the above sections (1–3) is in the upper half of the zone.

4. On the south side of the Harpsden Valley, at about half a mile west of Harpsden Court, a pit shows over 30 ft. of firm white chalk with many bands of solid and carious flint, and seams and veins of tabular flint. Mr. H. L. Hawkins has found an example of *Echinocorys scutatus* (ovate form) here; but megascopic fossils generally are very scarce. Bryozoa are abundant. This chalk must be near the base of the *M. coranguinum* zone.

5. Some 70 or 80 ft. of beds—the lowest of which may be represented in the upper part of the last section—are exposed in the new road-cutting in Harpsden Wood, between 1 and 3 furlongs west of Harpsden Court. The chalk is white and blocky, and contains numerous courses of solid nodular flints, and a few thin seams of the tabular variety. Fine marly partings are common near the middle of the section. At the lower or eastern end of the cutting there is a thin seam of marl which is full of remains of Asteroids, *Cidaris, Inoceramus*, and free Bryozoa (notably *Entalophora echinata* (F. A. Roemer). Three-and-a-half feet above the marl-seam is a band of rather hard, iron-stained, lumpy chalk, with crooked borings which are sometimes occupied by casts of a sponge. An *Echinocorys scutatus,* of a somewhat gibbous form, was seen about 20 ft. higher. *Inoceramus cuvieri* is the only common big fossil above the marl-seam.

In the highest part of the cutting (just within the area of Sheet 268) the Chalk is much piped by gravel.

6. Beside the high road on White Hill, east of Henley, at a point about one mile south of Remenham Church, there is a quarry with a vertical face of about 60 ft. The chalk is very flinty, and has a gentle dip to the south-east. In the lower part of the section there is a conspicuous marl-seam, with another 7½ ft. above it, and a thin band of hard yellow nodules 3 in. below. Beneath the yellow nodules the chalk contains very regular layers of thin tabular flint, and many small globular flints, including *Plinthosellæ.* Above that horizon the flints are mostly solid nodules of medium size, and, in places, very closely set. The lower marl-seam is very rich in Asteroid ossicles, remains of *Cidaris*, and broken Bryozoa.

The higher beds may be inspected on the opposite side of the road cutting. They contain little but broken *Inoceramus*, and *Porosphæra.*

7. Close to the same road, at points a little to the west of the second milestone out of Henley, and near Middle Culham Farm, there are two small pits in white chalk with regular courses of solid flint—some showing an agate-like banding. From the excavation on the north side of the road I obtained a cast of a large *Pleurotomaria perspectiva* ? (Mant.). The other pit—on the south side of the road and at a slightly lower level—contains more flints and fewer fossils. These pits are in a chalk younger than any seen in the last section.

8. An opening in beds which seem to be in the lower half of the zone occurs at a point nearly a quarter of a mile south-south-west of Rose Hill. This shows about 20 ft. of white chalk, containing a marl-seam and many courses of solid and carious flints, and seams of tabular flint. *Inoceramus* is very abundant. I have seen two examples of *Echinocorys scutatus* (ovate form) here.

9. Rather soft chalk with large solid flints is quarried (for whiting) in galleries beneath the Eocene cover east of Warren Row. The junction with the Reading Beds is well shown at the entrance to the workings. This chalk is probably the newest in the Henley district.

10. About half a mile north-east of Pudder's Farm, near Warren Row, a pit by the side of a lane shows over 20 ft. of flinty chalk which is probably rather above the horizon of that seen in pit (8). The lowest beds are of firm and widely-jointed character: the rest are softer and blocky. There is a dip of 3° or 4° to the south-east. Mr. Ll. Treacher has found well-preserved tests of *Micraster coranguinum* (Leske) here. Large fragments of thick *Inoceramus*-shell are very plentiful.

11. Flinty chalk is exposed in several places in the high bank on the

eastern side of the terrace-road running from Henley south-eastward to Wargrave (Sheet 268). In the clearest section, which is about 50 yards north-west of Park Place Bridge (west of Temple Combe), the bed of yellow nodules and overlying marl-seams seen in the quarry numbered (6) are recognisable. The lower marl-seam is less fossiliferous here. *Pharetrospongia strahani* Sollas is a common fossil in the chalk between the marls in both places.

Other sections of the Coranguinum Chalk are to be seen at Wheeler End Common, Moor End Common (swallow-hole), and to the east of Rockwell End, near the eastern border of the district; and in Hollandridge Lane, Nettlebed Wood, south of Wyfold Court, and near Hook End and Woodcote, farther west.

The highest beds of the zone, which are characterised by *Conulus albogalerus* Leske, seem to be unrepresented in this district, though they come in near the boundary of the Eocene main mass, a little farther to the south and south-east.

NOTE.—The Bryozoa recorded in the notes on the Chalk contributed by me to this Memoir were named—and, in many cases, also found—by Mr. Ll. Treacher, in whose collection they are preserved. I am indebted to him and to Mr. H. L. Hawkins for information relative to the fauna in many of the pits above described. Acknowledgment of the details kindly furnished by Messrs. C. P. Chatwin and T. H. Withers has already been made. H. J. O. W.

## LIST OF FOSSILS FROM THE MIDDLE AND UPPER CHALK.

The fossils from which the following list has been compiled were mostly obtained by Mr. Osborne White, and these are in his own possession; some, however, were found by Messrs. Ll. Treacher, H. L. Hawkins, C. P. Chatwin, and T. H. Withers, and are in their respective collections. Mr. J. Rhodes collected for the Geological Survey in 1887 from the exposures of the zone of *Holaster planus* on Aston and Chinnor Hills, and the fossils which he obtained are in the Survey Museum at Jermyn Street, London. The species so obtained are marked by an asterisk after the letters in the list.

The species entered in the column for the zone of *Terebratulina* were obtained by Mr. O. White from the upper beds of the zone when endeavouring to determine its upper limit.

In the column for the zone of *Hol. planus* the fossils found in the Chalk Rock (CR) have been separated from those which occur in the overlying "Micraster Beds" (MB), the letter *a* indicating occurrences on the escarpment (Ewelme, Aston and Chinnor Hills), while *b* indicates those in the inlying area around and north of Henley.

The numbers in the last column refer to those prefixed to localities in the zone of *M. coranguinum*, described in the text (pp. 51 to 52). A. J. J.-B.

| | Middle Chalk. | | Upper Chalk. | | | |
|---|---|---|---|---|---|---|
| | Zone of Rhynchonella cuvieri. | Zone of Terebratulina. | Zone of Holaster planus. CR | Zone of Holaster planus. MB | Zone of M. cortestudinarium. | Zone of M. coranguinum. |
| *PISCES.* | | | | | | |
| Corax sp. | – | – | – | b | – | 5 |
| Lamna appendiculata (*Ag.*) | – | – | a* | – | × | |
| Oxyrhina sp. | – | – | – | a | | |
| Ptychodus mammillaris ? *Ag.* | – | – | – | – | × | |
| *CRUSTACEA.* | | | | | | |
| Scalpellum fossula ? *Darwin* | – | × | | | | |
| *CEPHALOPODA.* | | | | | | |
| Baculites bohemicus (*Fritsch & Schloenb.*) | – | – | b | | | |
| Crioceras ellipticum (*Mant.*) | – | – | a b | | | |
| Heteroceras reussianum (*d'Orb.*) | – | – | a b | | | |
| „ sp. | – | – | a* | | | |
| Nautilus sp. | – | – | a | a | | |
| Pachydiscus peramplus (*Mant.*) | – | – | a* | | | |
| „ *cf.* peramplus | – | – | a* | – | | |
| Prionotropis woolgari (*Mant.*) | – | × | – | | | |
| Scaphites geinitzi *d'Orb.* | – | – | a b | | | |
| *GASTEROPODA.* | | | | | | |
| Avellana humboldti *Muller* | – | – | a | | | |
| Cerithium saundersi *Woods* | – | – | a* b | | | |
| „ cuckhamsliense *Woods* | – | | a b | | | |
| „ sp. | – | – | a | | | |
| Dentalium turoniense *Woods* | – | – | a* | | | |
| Emarginula (cast of) | – | – | a* | | | |
| Globiconcha sp. | – | – | a* | | | |
| Lampusia ? | – | – | a* | | | |
| Nerinea sp. | – | – | a* | | | |
| Pleurotomaria (Leptomaria) perspectiva (*Mant.*) | – | – | a* b | a | × | |
| Solariella gemmata (*J. de C. Sow.*) | – | – | a* b | a* | | |
| Trochus schlüteri *Woods* | – | – | a* | | | |
| „ berocscirensis *Woods* | – | – | a* | | | |
| Turbo geinitzi *Woods* | – | – | a* b | | | |
| *LAMELLIBRANCHIATA.* | | | | | | |
| Arca *cf.* galliennei *d'Orb.* | – | – | b | | | |
| Barbatia *cf.* geinitzi (*Reuss*) | – | | a* | | | |
| Cardita cancellata *Woods* | – | | a* | | | |
| Cuspidaria caudata (*Nilss.*) | – | | b | | | |
| Dimyodon nilssoni (*Hag.*) | | × | – | a b | – | 4, 5 |

| | Middle Chalk. | | Upper Chalk. | | | |
|---|---|---|---|---|---|---|
| | Zone of Rhynchonella cuvieri. | Zone of Terebratulina. | Zone of Holaster planus. CR | Zone of Holaster planus. MB | Zone of M. cortestudinarium. | Zone of M. coranguinum. |
| Inoceramus cuvieri *J. de C. Sow.* | – | × | – | a b | × | passim. |
| ,, involutus *J. de C. Sow.* | – | – | – | – | × | |
| ,, mytiloides *Mant.* | × | – | – | | | |
| ,, brongniarti *J. de C. Sow.* | – | × | | | | |
| ,, sp. (small) | – | × | a* b | a b | × | 9 |
| ,, sp. (like mytiloides) | – | × | a b | | | |
| Lima (Plagiostoma) hoperi (*Mant.*) | – | – | – | a | × | 5 |
| Nucula sp. | – | – | a* | | | |
| Ostrea hippopodium *Nilss.* | – | – | – | b | × | 1 |
| ,, normaniana *d'Orb.* | – | – | a | a* | | |
| ,, vesicularis *Lam.* | – | × | a b | a* b | – | 6, 7, 9, 10 |
| ,, sp. | – | × | – | b | – | 2, 4, 5 |
| Plicatula barroisi *Peron* | – | × | – | b | | |
| Pecten (Chlamys) cretosus *Defr.* | – | – | – | – | × | 6, 8 |
| Septifer lineatus (J. de C, Sow.) | – | – | a | | | |
| Spondylus latus (*J. Sow.*) | – | × | b | b | × | 4, 10 |
| ,, spinosus (*J. Sow.*) | – | × | a* b | a* b | × | |
| Trapezium trapezoidale (*F. A. Roem.*) | – | – | b | | | |
| *BRACHIOPODA.* | | | | | | |
| Kingena lima (*Defr.*) | – | – | – | – | × | 1, 6, 3 |
| Rhynchonella cuvieri *d'Orb.* | × | – | – | – | | |
| ,, limbata (Schloth.) | – | – | – | – | × | |
| ,, plicatilis (*J. Sow.*) | – | × | a* b | a b | – | 1, 5 |
| ,, reedensis *Eth.* | – | – | a* b | a b | × | 8, 10 |
| Terebratula carnea *J. Sow.* | – | × | a b | a b | × | 6 |
| ,, semiglobosa *J. Sow.* | – | × | a* b | a* b | × | 9 |
| Terebratulina gracilis *var.* lata *Eth.* | – | × | | | | |
| ,, striata *Dav.* | – | × | | b | × | 1, 3, 4, 7, 10, 11 |
| Thecidium wetherelli *Morris* | – | – | – | – | × | 6, 9 |
| *BRYOZOA.* | | | | | | |
| Berenicea regularis (*d'Orb.*) | – | – | – | – | – | 4 |
| ,, contracta *Seeley* | – | – | – | a | | |
| ,, disciformis (*Hag.*) | – | – | – | a | | |
| ,, papillosa, *var.* echinata *d'Orb.* | – | – | – | a* | | |
| ,, ,, *var.* grandis *d'Orb.* | – | – | – | a* | | |
| ,, phlyctænosa *Reuss* | – | – | – | a | | |
| ,, sp. | – | – | – | b | × | |
| Clausa francqana (*d'Orb.*) | – | – | – | – | – | 4, 9 |
| ,, globulosa (*d'Orb.*), *var.* crassicaulata *Greg.* | – | | – | – | – | Greenmoor Hill |
| Clinopora lineata (*Beissel*) | – | – | | – | – | 4 |
| Crisina cenomana *d'Orb.* | – | – | – | – | – | 5 |
| Crisina (Reticrisina) obliqua (*d'Orb.*) | – | – | – | – | | 5 |
| ,, ,, papyracea (*d'Orb.*) | – | – | – | – | | 4 |
| Entalophora echinata (*Roem.*) | – | – | – | – | – | 3, 4, 5, 6, 11 |
| ,, madreporacea (*Goldf.*) | – | – | | – | – | 1 |
| ,, virgula (*Hag.*) | – | – | | – | × | 4 |

| | Middle Chalk. | | Upper Chalk. | | | |
|---|---|---|---|---|---|---|
| | Zone of Rhynchonella cuvieri. | Zone of Terebratulina. | Zone of Holaster planus. CR | Zone of Holaster planus. MB | Zone of M. cortestudinarium. | Zone of M. coranguinum. |
| Homœosolen aculeatus *d'Orb.* | – | – | – | – | × | |
| Idmonea alipes *Greg.* | – | × | – | – | × | 10 |
| Meliceritites lonsdalei *Greg.* | – | – | – | – | – | 6 |
| Nodelea durobrivensis *Greg.* | – | – | – | – | × | 3, 4 |
| Petalopora costata *d'Orb.* | – | – | – | – | – | 4, 5 |
| Proboscina angustata *d'Orb.* | – | – | – | a | | |
| „ crassa (*Roem.*), *var* elevata | – | – | – | b | | |
| „ elevata (*d'Orb.*) | – | – | – | – | – | 9 |
| „ fasciculata (*Reuss*) | – | – | – | – | × | |
| „ radiolitorum (*d'Orb.*) | – | – | – | – | × | 4 |
| Reptoceritites rowei *Greg.* | – | – | – | – | × | |
| Reptomultisparsa sp. | – | – | – | – | × | 2, 10 |
| Siphoniotyphlus tenuis (*Hag.*) | – | – | – | – | × | |
| Stomatopora gracilis (*M. Edw.*) | – | – | a | ab | × | |
| „ granulata (*M. Edw.*) | – | – | b | a | – | 4 |
| „ spicea *Greg.* | – | – | – | – | – | 4 |
| Cribrilina sp. | – | – | – | – | × | 4 |
| Eschara lamarcki *Hag.* | – | × | | | | |
| „ sp. | – | – | – | – | × | |
| Hippothoa simplex *d'Orb.* | – | – | – | – | – | 4 |
| Membranipora dentata (*d'Orb.*) | – | – | – | – | – | 4 |
| „ elliptica *Reuss* | – | × | – | – | × | 5, 6, 7 |
| „ navicularis (*Marss.*) | – | – | – | – | × | |
| Micropora hippocrepis (*Goldf.*) | – | × | – | – | × | |
| Onychocella depressa ? *Hag.* | – | – | – | – | × | 5 |
| Tervia subgracilis (*d'Orb.*) | – | | – | – | – | 4, 5 |
| Vincularia sp. | – | | – | – | – | 9 |
| *ANNELIDA.* | | | | | | |
| Serpula ampullacea *J. de C. Sow.* | – | – | a* | – | | |
| „ cincta *Goldf.* | – | – | – | a* b | × | |
| „ fluctuata *J. de C. Sow.* | – | – | a | a b | × | 4 |
| „ granulata *J. de C. Sow.* | – | – | b | a* | × | 10 |
| „ ilium *Goldf.* | – | – | – | b | × | 1, 6, 9, 11 |
| „ pentangulata *S. Woodw.* | – | – | – | – | × | |
| „ plana *S. Woodw.* | – | – | a | a b | × | 9 |
| „ turbinella *J. de C. Sow.* | – | – | – | – | – | 10 |
| *ECHINODERMATA.* | | | | | | |
| Asteroidea (ossicles of) | – | × | b | b | × | passim. |
| Bourgueticrinus ellipticus (*Miller*) | – | – | – | b | × | passim. |
| Cardiaster cotteauanus *d'Orb.* | – | – | – | a | | |
| Cidaris clavigera *Koenig* | – | – | – | – | × | 3, 5, 6, 7, 9, 11 |
| „ hirudo *Sorignet* | – | | – | a* | × | 1, 3, 4, 5, 6, 9 |
| „ perornata *Forbes* | – | – | – | – | – | 1, 3, 6, 11 |
| „ sceptrifera *Mant.* | – | – | – | b | × | 1, 4, 5, 6, 9, 11 |
| „ serrifera *Forbes* | – | | – | – | × | 6, 9 ? |

| | Middle Chalk. | | Upper Chalk. | | | |
|---|---|---|---|---|---|---|
| | Zone of Rhynchonella cuvieri. | Zone of Terebratulina. | Zone of Holaster planus. CR | Zone of Holaster planus. MB | Zone of M. cortestudinarium. | Zone of M. coranguinum. |
| Conulus subrotundus *Mant.* | × | × | | | | |
| Cyphosoma radiatum *Sorignet* | – | × | – | b | | |
| „ sp. | – | – | – | – | – | 6, 7, 9 |
| Discoidea dixoni *Forbes* [= D. minima *Ag.*] | – | × | a | | | |
| Echinocorys scutatus *Leske* (gibbous form) | – | – | a* b | a* b | × | 5 |
| Echinocorys scutatus *Leske* (ovate form) | – | – | – | a b | × | 1, 2, 4, 8 |
| Helicodiadema fragile (*Wiltsh.*) | – | – | – | – | × | 4, 5, 6, 11 |
| Hemiaster minimus (*Ag.*) | – | × | | | | |
| Holaster placenta *Ag.* | – | – | – | a b | × | |
| „ planus (*Mant.*) | – | – | a* | a b | × | |
| Metopaster parkinsoni (*Forbes*) | – | – | – | b | × | 6 |
| Micraster coranguinum (*Leske*) | – | – | – | | – | 1, 3, 6, 7, 8, 9, 10, 11 |
| „ corbovis *Forbes* | – | × | – | b | | |
| „ cortestudinarium (*Goldf.*) | – | – | – | a b | × | |
| „ leskei *Desmoulins* | – | – | a* | a* | | |
| „ præcursor *Rowe* | | – | a* b | a* b | × | 4 ? |
| „ sp. | – | – | a b | | | |
| Mitraster hunteri (*Forbes*) | – | – | – | – | – | 1 |
| Nymphaster coombei (*Forbes*) | × | | | | | |
| Oreaster pistilliferus (*Forbes*) | – | – | – | – | – | 9 |
| Pentacrinus sp. | – | – | – | b | × | 6, 9 |
| *ANTHOZOA.* | | | | | | |
| Micrabacia coronula (*Goldf.*) | – | – | a* | | | |
| Parasmilia centralis (*Mant.*) | – | – | a* b | a b | | |
| „ granulata *Duncan* | – | – | a* | | | |
| *SPONGIÆ.* | | | | | | |
| Cliona cretacea (*Portlock*) | – | – | a | | | |
| Coscinopora infundibuliformis *Goldf.* | – | – | a | | | |
| „ quincuncialis (*T. Smith*) | – | – | – | b | × | 1 |
| Doryderma ramosum (*Mant.*) | – | | – | – | × | |
| Guettardia stellata *Mich.* | – | – | a* | | | |
| Pharetrospongia strahani *Sollas* | – | | – | a b | × | 1, 6, 7 |
| Placotrema cretaceum *Hinde* | – | – | a* | | | |
| Plinthosella squamosa Zitt. | – | – | – | – | × | 1, 6, 7, 8 |
| Plocoscyphia convoluta (*T. Smith*) | – | – | a* b | a* b | × | |
| „ fenestrata (*T. Smith*) | – | – | a* | | | |
| Porosphæra globularis (*Phill.*) | – | × | – | a b | × | passim. |
| „ patelliformis *Hinde* | – | – | – | – | – | 1, 5, 9 |
| „ pileolus *Lam.* | – | – | – | – | – | 5, 6 |
| „ sp. (branching) | – | – | – | – | – | 6, 7 |
| Ventriculites alcyonoides *Mant.* | – | – | – | b | | |
| „ impressus ? (*T. Smith*) | – | – | a* | a | × | |
| „ infundibuliformis *S. Woodw.* | – | – | – | a b | | |
| Ventriculites mammillaris *T. Smith* | – | × | – | a b | × | |
| „ radiatus *Mant.* | – | – | a | a* b | | |

# CHAPTER VIII.

## EOCENE.

By H. J. Osborne White.

Strata of Eocene age occupy a very small area of the country represented on Sheet 254, and occur only as outliers, scattered over the dipward slope of the Chalk uplands, in the eastern and southern parts of the district.

The thicker of these residual masses form low, gorse-grown or wooded hills, of mound-like contour, which contrast with the more even and gently-inclined surfaces of their supporting ridges of chalk, and may often be recognised from a considerable distance. In addition to the broken, hummocky character of their sides, indicative of small landslips, many of them show signs of greater disturbances—usually shallow synclinal flexings and normal faultings—the effect of which has been to place their constituent rocks somewhat below the normal position, and thus to render them less accessible to the agents of detrition. The common occurrence of Eocene outliers in places where there is a slight change in the angle or direction of the dip, on the northern side of the London Basin, was noticed, many years ago, by Mr. W. Whitaker, who also drew attention to the distribution of these relics along lines or belts of country roughly parallel to the boundary of the Eocene main mass.[1]

On the south-west, in the vicinity of the Thames at Goring and Cleeve, Eocene Beds are found on the crest of the Chilterns, at the edge of the Chalk escarpment; but on the north-east, near Stokenchurch, they have been stripped off the Chalk for some distance down the dip-slope, and the Chalk itself has been subjected to a sort of bevelling, which has removed much of its Upper division from the summit of the upland—there 200 ft. higher. The little outliers formerly mapped near the crest-line and even on the scarp-face of the Chiltern Hills, north-eastward of the village of Nuffield, are merely masses of reconstructed material in the Clay-with-flints, and the contents of truncated pipes, lying well below the basal plane of the Eocene strata.

The older Tertiary Rocks recognised in this area are:—

2. London Clay.

1. Reading Beds.

1 'The Geology of the London Basin' (1872), pp. 351-352; see also Whitaker 'On Subaërial Denudation, and on Cliffs and Escarpments, &c.,' *Geol. Mag.*, vol. iv (1867), p. 487.

## Reading Beds.

The Reading Series here has much the same character as at the type-locality in the adjoining district on the south—that is to say, it consists of irregular and impersistent beds of light-coloured, current-bedded sand, and of mottled plastic clay, with subordinate lenses or seams of fireclay, loam, and flint-pebbles. At Lane End, and probably also at Stoke Row, however, these sediments are supplemented by quartzose gravels and gravelly sands of fluviatile aspect.

The lowest, or "Bottom" bed, is, as usual, a roughly-laminated dark grey clay, with some speckled greenish sand and loam containing green-stained flint-nodules, imperfectly-rounded small flint-pebbles (also green-coated), and occasional layers of oyster-shells. The "leaf-beds"—seams of fine, pale-tinted clay, with bits of lignite and impressions of leaves—so well seen at Reading, have been observed, in their accustomed place near the base of the series, at Lane End.

In the few places where the Bottom Bed is clearly exposed its junction with the Chalk is even and sensibly parallel to the courses of flint-nodules in that rock. There is, nevertheless, a considerable unconformity between the Lower Eocene and Upper Cretaceous in this part (as in other parts) of the country; for higher zones of the Chalk than are here represented have been recognised at no great distance to the south-west and south-east; and even within the limits of the Henley district, where the Reading Beds are everywhere in contact with the thick and hardly-divisible zone of *Micraster coranguinum*, a comparison of the faunal and lithological features of the chalks which form the Eocene platform in different places gives ground for the belief, that the Lower Eocene Beds gradually overstep the Upper Cretaceous, towards the north or north-west.

Where fully developed, the Reading Beds attain a thickness of from 50 ft. to 60 ft.—a thickness not uncommon in the north-western part of the London Basin.

In dealing with the several patches of this formation it will be well to commence with the interesting outlier of Lane End, wherein the many phases of the Reading Beds are best exemplified, and the disturbances mentioned at the beginning of this chapter have their strongest expression.

### *Lane End.*

The following account embodies the greater part of that given by Mr. Whitaker in "The Geology of the London Basin,"[1] and some more recent notes by the late Mr. J. H. Blake and by the present writer.

"This strangely-shaped mass of the Reading Beds radiates in three directions from a central part near Lane End—on the

[1] *Mem. Geol. Surv.*, vol. iv, 1872. The description of the Lane End outlier there given is taken from an earlier work on 'The Geology of Parts of Middlesex, Hertfordshire, Buckinghamshire, Berkshire, and Surrey' (*Explan of Sheet* 7, *Old Series*), 1864, pp. 31 to 34.

N.W. over Cadmore End Common; on the S.S.W. to Moor Common, where the beds are much disturbed; and on the E. to Park Hill, and is capped in many places by London Clay. . . . Each of the three spurs is much broader towards the part away from the centre; so that the outlier is, as it were, made up of three masses joined on by narrow strips to a fourth and central mass."[1]

The north-western part of the outlier, around Cadmore End Common, consists, for the most part, of sand. Mr. J. H. Blake found evidence there of two faults which are not shown on the map. "One in the northern part [of the Common] runs in an easterly direction, having a downthrow of the Reading Beds to the north, with the Chalk at the surface to the south; another trends in a south-westerly direction with a downthrow of the Reading Beds to the west. . . . At the brickyard at Kiln Wood, on the west side of Cadmore End Common, sections of the Reading Beds have been exposed, showing mottled clays overlying from 16 to 18 ft. of white sand, in which latter impersistent bands of bluish-black sand occurred, varying from two to four inches in thickness." A sample of this dark material, taken by Mr. H. B. Woodward, was examined chemically by Dr. W. Pollard, who reported as follows:

"The specimen treated with strong hydrochloric acid leaves an insoluble residue of quartz and a little clayey matter. The solution contains alumina, iron, manganese (as peroxide), cobalt, and nickel. Small quantities of lime and magnesia are also present."

The occurrence of cobalt and nickel in the Reading Beds had not previously been recorded.[2]

A little to the west of Bolter End, on the south-western side of the high road, there is an oval hillock, about one furlong in longest diameter, which seems to consist entirely of pebbles in a scanty matrix of yellow, ferruginous sand. The deposit, which recalls the Blackheath Beds to the south and south-east of London, is worked to a depth of 12 ft. (formerly 16 ft.) in a pit at the south-eastern end of the hillock, and, to judge from surface-indications, has a maximum thickness of about 20 ft. A current-bedded structure is indicated by the gradation in the size of the pebbles, which are all of flint. This bank of shingle lies in the lower part of the Reading Series; its base, at the south-eastern end, being probably 10 to 12 ft. above the Chalk.

About 100 yds. to the south-west of the ("Peacock") Inn at Bolter End cross-roads, part of the Bottom Bed, presenting some unusual features, is shown in a water-hole and hedge-bank on the north-west side of the Fingest road. The bed, of which only a few inches are exposed, is a light brown loamy marl, containing, or filling up spaces between, angular blocks and

1 'Geology of the London Basin,' p. 207.
2 J. H. Blake, 'Summary of Progress of the Geological Survey for 1900' (1901), pp. 122-123.

more or less rounded pieces of (*a*) firm to very hard, compact, cream-coloured chalk, having a conchoidal fracture, and of (*b*) hard, rough, grey limestone, with vague brown inclusions. The marl consists to a large extent of minute plates and lumps of calcite (which resemble the coarse residues obtained on washing samples of the subjacent Chalk), together with fine quartz-sand and some argillaceous matter. A small pipe-like hollow, or pocket, in the Chalk, exposed in the road-bank close by, and at a slightly lower level, is lined with a fine mealy greyish-yellow sand, composed of these little calcite chips, with a variable but small proportion of fine quartz-grains.[1] The sand also penetrates into the joint fissures of the adjoining undisturbed Chalk, and has yielded, besides numerous small sub-cylindrical concretions, some fragments of *Inoceramus*, *Entalophora virgula* (Hag.), *Helicodiadema fragile* (Wilt.), and *Micraster coranguinum* (Leske), derived from that rock.

Thin sections of specimens of the hard cream-coloured chalk (*a*), and of the grey stone (*b*), occurring in the Reading Bottom Bed, have been examined microscopically in the laboratory of the Geological Survey by Mr. H. H. Thomas, who supplies the following notes:—

"*a* (Laboratory number, E. 4928.) A fine-grained chalk, with no signs of re-crystallisation, and very free from detrital minerals of recognisable dimensions. It is made up almost entirely of complete and broken bits of foraminifera, with a very few larger shell-fragments."

"*b* (Laboratory number, E. 4929.) Chalk; almost completely re-crystallised into a mass of very fine granular calcite. It is remarkable for showing patches, or rather grains, of quartz ranging up to 0·45 m.m. in diameter. That these grains are original is shown by the fact that rows of inclusions end abruptly at the margin of the grain.[2] These quartz grains are scattered through the chalk at very frequent intervals, and are not arranged in lines. No organic remains can be recognised (except a fragment or two of larger shells), the foraminifera having been destroyed by the re-crystallisation. A small quantity of iron ore, in the form of limonite, is present, staining the chalk locally."

At the northern end of the brickyard, a quarter of a mile north-west of the cross-roads in Lane End, Mr. Blake saw 9 ft. of slightly loamy sand, brown in the upper part, white and buff in the middle, and ash-coloured at the bottom. At the southern end of the same working there was 6 ft. of grey and crimson clay, underlain by yellow and brown sand. In another part of the yard grey mottled clay, with bands of sand, hardened in places, was noted, and, beneath it, laminated clay with leaves, and the Bottom Bed, with sharks' teeth and oysters, were said to have been encountered. At or near this spot the "leaf-beds" were observed also by Mr. R. H. Tiddeman, in 1868.[3] Small exposures of sand and mottled clay occur in the eastern part of the outlier, round the hill south of Park Lane.

Less than 100 yards to the south-east of the church (Holy Trinity) at

1 A somewhat similar deposit was observed by Mr. Clement Reid at Rye Hill, near Bere Regis, in Dorsetshire. See 'The Geology of the Country Around Dorchester,' 1899, p. 18. Mr. Reid remarks that these pipes "are obviously contemporaneous with the Reading Beds, and not due to subsequent percolation of rain-water, which would have destroyed the delicate shells of ostracoda and foraminifera" present in the sand at that spot.

2 Quartz-grains of detrital origin, however, are not wanting. [H. J. O. W.]

3 'Geology of the London Basin,' p. 207.

Lane End, and close to the road to Moor End Common and Frieth, a sand-pit showed (1906) the following remarkable section [1] (Fig. 5):—

| | | Ft. |
|---|---|---|
| London Clay. | 1. Soil and made ground, mostly of brown clay. | |
| | 2. *Basement Bed.*—Evenly-stratified brown loamy sand, with continuous seams of light blue-grey clay, and of dull green, speckled, glauconitic sand: small concretions and laminæ of brown clay ironstone. At the base, a slightly undulate band (6 in. thick) of glauconitic loamy sand, containing flint-pebbles (some large) and a few small quartz-pebbles - 5 to | 6 |
| Reading Beds | 3. Buff and brown sand, loamy in the upper part, with some small quartz-pebbles and seams of light grey clay. (This division is cut out by the London Clay at the northern and southern ends of the section, and its base, which is marked by a thin but persistent band of clay, curves downward in the middle of the section so as to truncate the irregular beds in the division below) - - - - - - - - - to | 3 |
| | 4. *a.* Buff and brown sand, with seams of grey clay, to 2 ft.<br>*b.* Lens of current-bedded quartz, flint, and clay gravel - - - - - - to 3 ft.<br>*c.* Fine buff sand, with small lenses of gravel, and a seam of laminated clay at the base - to 2 ft.<br>(In the southern part of the section, beds (*a*), (*b*), and (*c*) appear to be represented by alternations of blue-grey clay and ferruginous sand)<br>*d.* Irregular beds of gravel, cemented in places—and especially in a band, a few inches wide, at the top—by red and brown iron oxide - to 4 ft.<br>*e.* Buff, yellow, and light brown gravelly sand, markedly current-bedded and full of small quartz-pebbles. A lens of gravel seen in one place - - - - - - to 5 ft. | 15 |
| | 5. Friable, coarse-grained sandstone, without pebbles: light brown above, and greenish-yellow below - - | 2 |
| | 6. *a.* Firm brown sand, with scattered small pebbles, 2 ft.<br>*b.* A bed or patch of wet gravel, just touched at the bottom of a temporary hole in the floor of the pit - - - - - - - seen, 6 in. | 2½ |

The coarser materials of the gravels and gravelly sands are—

*Quartz-pebbles:* white, yellowish, and pink; translucent and opaque. These are generally incompletely rounded, and range up to 1½ inches in diameter. They form the bulk of the gravel.

*Sub-angular flints:* mostly bleached a dull white, but commonly possessing a green tinge on the outside. Many are red within. A large proportion of the flints in this condition are fragments of tabular seams or veins; and pieces of "carious" or "cavernous" nodules are abundant. Some of the less worn flints, which have formed part of solid nodules, fall into small splinters when removed from their matrix. Size:—common up to 2 in.; a few ranging up to 6 or 7 in. in diameter.

*Flint-nodules:* little worn. Mostly small and of globular and sub-cylindrical forms. Some are hollow and spongeous, and many have the dark green coating and corroded surface which characterise the nodules in the Reading Bottom Bed. Small club-shaped nodules, enclosing portions of stems of Coscinoporæ, are fairly common.

*Flint-pebbles:* black and grey; fairly common up to 1½ in. in diameter, but greatly outnumbered by sub-angular flints.

*Clay:* in sub-angular pieces and pebbles up to 4 in. in diameter. In

[1] An account of this section appeared in *Proc. Geol. Assoc.*, vol. xix, 1906 p. 371.

*Fig. 5.—Section of Sand-pit South-east of Lane End Church.*

View of Eastern side.

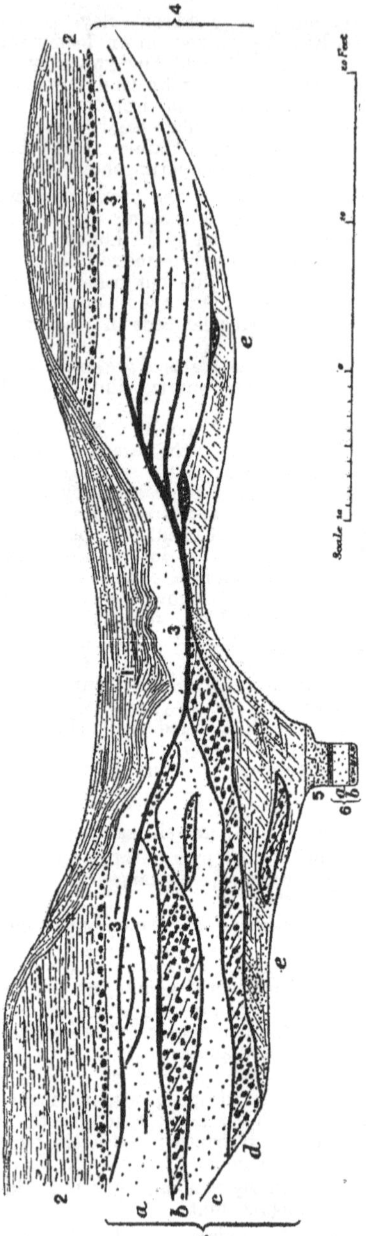


1. Soil and Made Ground.
2. Basement Bed of London Clay.

3–6. Reading Beds.

(For detailed references, see p. 62.)

some places these are so closely set as to give the gravel the appearance of a breccia.

*Ironstone:* in flaggy pieces, and in concretions enclosing pebbles of clay and other material.

*Lydite:* black, polished, imperfectly rounded pebbles of chert or other siliceous rock, often dull brown within. Mostly under ½ in. in diameter. These closely resemble the pebbles occurring at the base of the Portland Beds in Oxon and Bucks.

*Quartzite,* or very compact fine-grained sandstone, of light grey and light brown tints: in small ovate pebbles, to ⅝ in. in longest diameter.

*Sandstone:* brown, in pebbles to about ½ in. in diameter.

*Silicified* Inoceramus-*shell:* small pieces abundant in places.

With the lithological characters of the Chalk in the Chiltern Hills and Berkshire Downs chiefly in view, one may say that the flints in these gravelly beds have an upper *M. cortestudinarium* and lower *M. coranguinum* Zone facies. So far as the writer is aware, the Lower Eocene strata on the northern slope of the London Basin nowhere rest upon chalk of greater age than that of the upper half of the *M. coranguinum* Zone, but the transgressive relations of the Reading Beds to the Chalk in that area render it highly probable that older zones of the latter formation were exposed to early Eocene erosion in the Midlands. A northern or north-western derivation is also to be inferred for the quartz-pebbles, which here attain a size, and occur with a profusion, unexampled in the main mass of the Reading formation, to the south and east.

The gravel beds are seen again in a sand-pit near the kiln, about 200 yds. east-south-east of Muzwell Farm (a former section at this spot is detailed below). Their mode of occurrence and general composition are the same as in the above excavation, but the average size of the quartz-pebbles is rather greater, and the tabular and spongeous flints are more abundant.

The inclined and faulted beds in the southern lobe of the outlier are thus described by Mr. Whitaker:—

"On the common S.W. of Lane End there are many pits in the sand of the Reading Beds, in some of them capped by the loamy basement-bed of the London Clay.

"Further in the same direction the Reading Beds continue down the slope of the hill, instead of ending off as one would expect, showing that they have been here affected by some disturbance.

"On the common below Stockwell [one furlong north-west of Moor Farm] both sand and plastic clay have been found, and it is possible that a spur of sand may stretch over to Moor Farm. At the eastern side of the Common a few feet of the Reading Beds were passed through in sinking for chalk, which is worked here by means of a shaft and underground galleries. In a sand-pit between the kiln and Mozzles[1] the following section was shown (1859):—

"Basement-bed of the London Clay. Brown clayey sand, with green grains and flint-pebbles, about 2 ft.

Reading Beds. Sand, chiefly white, with layers of laminated iron-sandstone; the upper part very coarse, the lower finer. I was told that 25 ft. of sand had been found.

[The "very coarse" sand mentioned here is probably that in which the quartzose gravel is now to be seen. The beds have a marked south-westward dip.]

[1] Muzwell Farm, on the New Series Map, Sheet 254.

"The southern part of the outlier, from Mozzels to Moors End, was very hard to make out, the beds being much disturbed, and thrown, by faults apparently, into places where one would not at all expect to find them. Strangely, too, it would have been almost impossible to map this part at all rightly but for the clear *natural* sections that occur, a very unusual thing in inland Tertiary-districts. These sections, however, were not to be seen without some trouble.

"The wood W.S.W. of the ruined church of Ackhamstead[1] does not look, from any spot outside it, a very promising place for geological sections, and the thick underwood would be quite enough to satisfy many explorers as to the need of forcing their way through it. However, the water that flows over the moist ground of the Common on the north collects, inside the wood, into streams, which, having a sharp fall, cut for themselves channels of some depth, often rather steep-sided and with miniature waterfalls. In one of these the London Clay and its basement-bed (with ironstone, septaria, and fossils) were cut into, showing that this formation occurs here at a comparatively low level, far lower than the Reading Beds at Lane End. The streams all flow into a large and deep swallow hole at the eastern end of the wood. This is in the Chalk; but the Reading Beds come on at once: for I saw the green bottom-bed in a saw-pit close by, at the southern edge of the wood. At Moors End there is mottled plastic clay by the roadside. Now, both that place and the saw-pit are in the bottom of the valley, whilst the rising ground to the S.E.—that is, in the direction of the general dip of the country (a direction, therefore, in which one would expect higher beds to come on) is bare Chalk, giving evidence of a fault, or other disturbance, running N.E. from Moors End and throwing down the Tertiary beds on the N.W.

"At the most southern point of Moor Common—that is, just above Moors End, I found traces of the basement-bed, and higher up, by the cross-roads, there were some small shallow pits in the London Clay itself. The hill to the west, however, is Chalk, capped by 'clay-with-flints' at the higher parts. This state of things continues further northward, for on the eastern side of the road that, skirting Moor Common, runs N.N.W. along the bottom of the valley, there are frequent signs of the Tertiary beds, whilst on the other side the Chalk rises to a much greater height than that of the London Clay on the Common. This gives evidence of a second fault, or other disturbance, running N.N.W. from a point a little west of Moors End to the bend of the road west of Mozzels, and with a downthrow on the east.

"There are swallow-holes at the western corner of the wood on the eastern side of the road at the above bend, and here again some of the streamlets that feed them lay bare the beds in the wood, and show along their channels sand (with pipe-clay and iron-sandstone) and mottled clay. Now, not only is the hill to the S.W. chalk, but so also is that to the north, leading one to conclude that there is a third fault, with a southerly downthrow, along the small valley N.W. of Mozzels.

"A stream runs along the line of the second and main fault at its northern end, and flows into the swallow-holes in the wood.

"On the western side of this fault a small patch of the mottled clay and the bottom-bed spreads a short way up the flank of the hill, along the road to Perrin's Spring [N. of Frieth], showing that there is here not a fault only, but also a trough.

"To return to the wood west of Ackhamstead: It is on the Tertiary beds and the London Clay comes very near to its north-eastern edge, whilst the rising ground close by, towards Moor Farm and above Ackhamstead, is Chalk, leading one to infer the presence of a fourth fault, or other disturbance, running N.W. along the edge of the wood to Mozzels; an inference which is strengthened by the fact that the two S.W. and N.E. faults described above seem to end suddenly on the N.E. (as they also do on the S.W.), as if cut off by another fault at right angles, or nearly so, and with a downthrow on the S.W.[2]

[1] This name is not engraved on Sheet 254. The spot is there marked by words "Church (Ruins)."

[2] The south-western fault is incorrectly printed, and the north-western fault omitted, in Sheet 254.

"I was led to think that these four disturbances are faults, by the sudden manner in which the Tertiary beds are brought on by them, as well as by the fact that they all seem to end off suddenly at both ends, as if cut off by other faults. To suppose an area to be altogether surrounded and included by faults, and dropped down in mass, may be thought an almost unwarrantable thing by very many geologists, and these four faults would not have been marked on the Geological Survey Map had the difficulties of the case seemed surmountable in any other way. This outlier was worked at carefully for a long time, and some time after it had been mapped, Sir A. Ramsay went to Lane End with me, and after a detailed examination of the ground and sections we were unable to suggest any other explanation. Fig. 6 illustrates the appearances above-described.

Fig. 6.—*Section across the Lane End Outlier.*[1]

(W. Whitaker.)

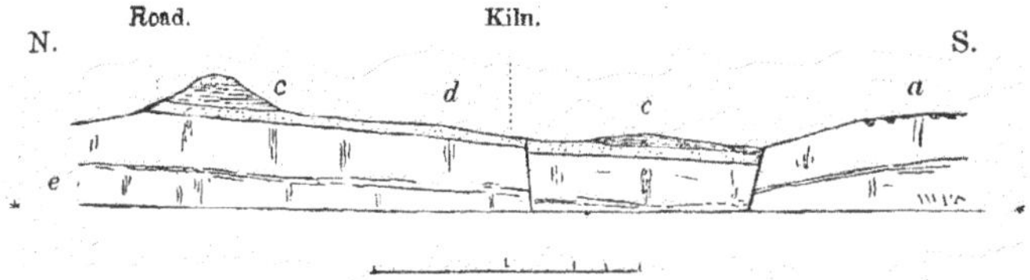


Horizontal Scale 2 in. to a mile. Vertical Scale twice as large

(*a*) Clay-with-flints. (*d*) Reading Beds.
(*b*) Pebble-gravel. (*e*) Chalk.
(*c*) London Clay. ** Approximate sea-level.

"From Mozzels northward to Bolter End the boundary-line is much hidden by wood."[2]

More recent observations support Mr. Whitaker's views respecting the structure of the southern branch of the Lane End Eocene outlier. The strata beneath the valley lying between the villages of Lane End and Frieth have been bent into a trough, and there is now little doubt that fracturing of the *graben* type has accompanied or followed upon the development of the flexure. In settling down, the fault-block has acquired a rather strong tilt towards the north-west; the vertical displacement (by faulting and flexing combined) at its lower end, near Muzwell Farm, being about 150 to 160 ft. A result—possibly a direct result—of this tilting has been the production of a local north-westward drainage; the valley coinciding with the faulted block sloping rather quickly in that direction, and against the prevailing dip of the rocks and inclination of the valleys in the surrounding country.

It should be mentioned that the swallow-hole at the eastern corner of Moor End Common (in Moor Copse) is an exceptionally good example of its kind; with vertical sides which show the junction of the Chalk and Reading Beds, and a sandy floor in which open fissures are sometimes to be seen. The broken state of the ground, and the inclined and prostrate trees, along

[1] Fig. 29 of the 'Geology of London,' p. 185.

[2] 'The Geology of the London Basin,' pp. 208-9; and 'Geology of London vol. i, pp. 184-5.

the courses of the little brooks which feed the swallow-hole, show that the wasting of the soft Eocene Beds is there proceeding apace; and the volume of sand and mud already carried down into the Chalk must be very large.

*Ibstone.*

The Reading sands and clays to the south of this place are either faulted or "piped" into the Chalk, for the crest of the narrow ridge in which they occur seems nowhere more than 170 ft. higher than the outcrop of the Chalk Rock on the steep slope to the south-east; and a pit 100 yds. west of Ibstone House, in the midst of the outlier, as mapped, showed Chalk covered only by Clay-with-flints.

In Park Wood, where the pit just mentioned is situated, Mr. Whitaker[1] saw an excavation in white and coloured soft sand with small patches of crimson and pale blue clay, and rounded flints in places.

*Turville Heath and South End.*

An outlier of the Reading Beds, for the most part thin and resting on an uneven, much-piped surface of Chalk, extends southward from Turville Heath, by Summer Heath and South End, to Bosmore, near Stonor.

The banks of the main road, and some old pits, south-east of Blundell's, at the northern end of the outlier, show a little disturbed sandy loam and brown clay.

Red-brown sandy clay occurs in pipes in the Chalk close to the boundary of the Reading Beds east of the church at Turville Heath; and yellow sand has been dug on the eastern side of the main road at the southern end of Summer Heath.

At the brick-works one furlong south-east of the ("Drover") Inn at South End there are shallow pits in brown sandy loam, which seems to be underlain by mottled clay and yellow sand. The Reading Beds here descend some distance below their proper level down the steep slope on the eastern side of the outlier, and on the old 1-in. map (Sheet 13) an east-and-west fault, with a downthrow to the north, is indicated. The broken character of the ground in and around the brickyard shows that the clays and sands are slipping into the steep-sided combe (of Binfield Bottom) which commences at this spot; but as the Chalk is close to the surface on the line of the road to the south of South End, and no evidence of the presence of Reading Beds above it was obtained short of the north-eastern end of the patch of Pebble Gravel near Kimble Farm, it is probable that there is some small dislocation, affecting both the Eocene rocks and the Chalk.

Farther south, more sandy beds come in, and the boundary of the outlier is better defined. Yellow and brown sand has been dug in small pits at the south-eastern corner of Stonor Park and at Cockslease Farm; while at Bosmore, where the outlier terminates in a hillock, the same material, containing seams of ironstone and flint-pebbles, was formerly worked to a depth of more than 20 ft., and may still be seen in a shallower excavation.

*Fawley.*

At Lower Woodend Farm, three-quarters of a mile north-north-east of Fawley Church, there is a little drift-capped outlier about one furlong in longest diameter—somewhat larger than is shown on the map. Small pits opened near the middle of the patch show 5 or 6 ft. of bright yellow sand, stained and, in places, concreted by iron oxide, and containing impersistent layers of black flint-pebbles.

*Russell's Water.*

A thin and ill-defined spread of loam and sand extends from Russell's Water, south-eastwards, for about 1½ miles, to Maidensgrove. No clear

[1] 'The Geology of the London Basin' (*Mem. Geol. Surv.*), vol. iv, 1872, p. 206.

sections of the beds here now exist. In 1858 Mr. Whitaker saw light brown loam or brick-earth, with small rounded and angular flints, overlying buff and white sand, by the brickyard (since disappeared) at Russell's Water. "The Chalk had not been reached after sinking 18 ft., whilst in making the kiln, close by and at a higher level, that rock was found to be only 8 ft. from the surface, and still higher up the hill it is free from Tertiary beds, as if there were some disturbance. On the Common there are many small pits, in which are variously-coloured sands and clays beneath the drift. There seems to be another disturbance in the valley on the north-east, as on that side the Reading Beds are found at a much lower level than to the south-west, and in a small chalk-pit there is a dip of 8° in a direction 25° east of north—*i.e.*, down the slope of the valley."[1] It is probably the slight local reduction of the general south-eastward dip, incidental to this north-north-eastward tilting, that gives the spur supporting the Russell's Water outlier its step- or terrace-like aspect in a distant view.

## *Nettlebed.*

Rising nearly 100 ft. above the general surface of the Chalk in its vicinity, this hill of Eocene sands and clays, with its broken, furzy slopes and crowning windmill, forms one of the most prominent and pleasing features of the Chilterns.

The Reading Beds here are between 50 and 60 ft. thick, and retain, in at least two places, a thin covering of London Clay. They have long been worked for bricks, tiles, and drain-pipes, and are well exposed in excavations on the southern and south-western sides of the hill. Nearly all the pits now or recently worked, however, are in the upper half of the series, and although some idea of the character of the lower beds can be obtained from small exposures scattered over the Common, their succession is known only from accounts published 40 or 50 years ago.

Sir Joseph Prestwich[2] made out the following section "from several small pits in [a] brickfield" at Nettlebed:—

| | | Ft. |
|---|---|---|
| "1. | Gravel, chiefly flint-pebbles—averages about - - - - | 5 |
| *e.* | White and yellow sands with ironstone nodules; flint and a few quartz pebbles occur in the lower part of this bed - - - - - - - - - | 12 |
| *d.* | Light greenish clay, whitish sand, and masses of sandstone, and red sands and clay - - - - - - | 10 |
| *c.* | Red / Black / Light green } slightly mottled clays - - - | 7 |
| *b.* | White sand with some large and very compact concretionary sandstones in the upper part of it, also a few dark red ferruginous nodules - - - - - - | 10 |
| *a.* | Green sand and flints - - - - - - - | 1 |
| | | 45 |
| 4. | *Chalk* - - - - - - - - - - - - | 10" |

No section resembling the above is now to be seen.

In the Geological Survey Memoir on Sheet 13 (Old Series),

[1] 'The Geology of Parts of Oxfordshire and Berkshire' (*Mem. Geol. Surv.*), 1861, p. 43.

[2] 'The Woolwich and Reading Series.' *Quart. Journ. Geol. Soc.*, vol. x, 1854, p. 89.

*Fig. 7.—Sections in the Nettlebed Outlier.*

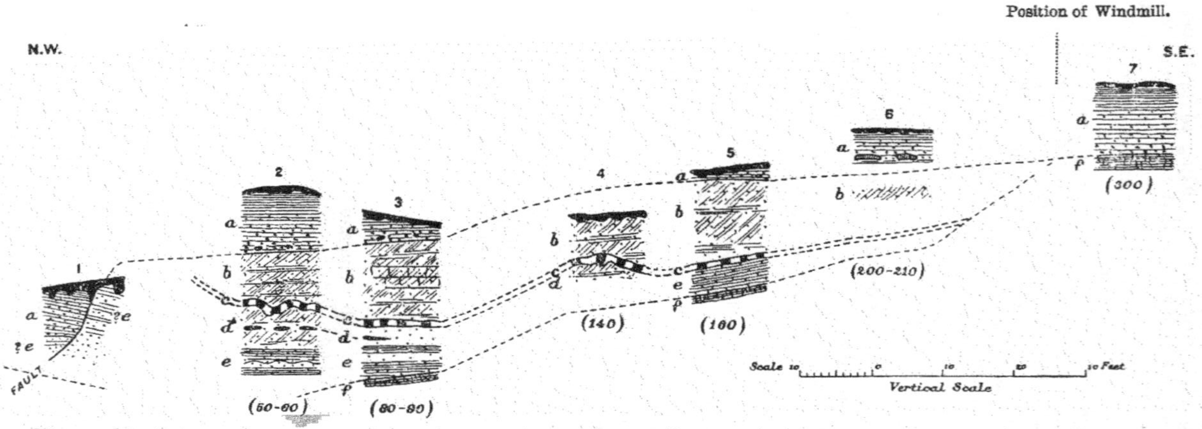


The numbers in parentheses, *e.g.* (140), denote approximate distance in yards of the several sections from the fault in section 1.

*a.* Base of London Clay : loam and sand, with pebbles and concretionary limestone.

| | |
|---|---|
| *b.* Current-bedded sand<br>*c.* Conglomerate bed<br>*d.* Current-bedded sand, with ironstone<br>*e.* Sands and loams<br>*f* Mottled plastic clay. | } Reading Beds. |

which was published in 1861, Mr. Whitaker gives[1] the following section, made up, like the last, from several pits:—

"Gravel.

London Clay and its basement bed, 20 to 25 ft.

Reading Beds.
1. Greenish clay, only seen at one place, and then not thick.
2. Brown, white, and crimson sands, several feet.
3. a. Plastic clay, variously coloured, with sandy clay and a little sand; several feet.
   b. White or very pale green sandy clay (fire-clay) and sand.
   c. Plastic and sandy clays, brown, blue and crimson.
4. Sands.
5. Bottom-bed; probably thin."

The Bottom-Bed, which was invisible at the date (1858) of Mr. Whitaker's examination, was seen, in more recent years, at Soundess Kiln (on the north-eastern slope of the hill) by Mr. J. H. Blake, who noted:—

| | | Ft. |
|---|---|---|
| Reading Beds. | Grey clay - - - - - - - | 3 |
| | Brown loam - - - - - - - - | ? |
| | Sand - - - - - - - - - | 1 |
| | Greenish-coloured sand with green-coated flints at base - - - - - - | 3 |
| Chalk. | | |

In the upper part of the hill, on its south-western side, a succession of pits now (1906) gives an interrupted section of the higher beds, about 300 yds. in length. The gaps are many and wide, but, thanks to the persistence of a bed of very distinctive character, the sub-divisions recognised in different exposures can, in most cases, be easily collated. The relative positions and the principal features of the clearest of these exposures are indicated in Fig. 7.

| Explanation:— | | Ft. |
|---|---|---|
| | Soil, of clay or loam, with pebbles. | |
| London Clay. | a. Evenly-bedded brown loam and loamy sand with seams of blue-grey clay and small ferruginous concretions: a thin and impersistent seam of flint-pebbles (some of large size) at the base. Masses of concretionary limestone in section (6) - - - to | 10 |
| Reading Beds. | b. Yellow and brown current-bedded sand; crimson patches in places. Partly hardened into friable sandstone in section (2) - to | 12 |
| | c. Conglomerate, consisting of pebbles and sub-angular pieces of coloured clay (to 1 ft. diameter) in coarse current-bedded sand, with streaks of clay and loam. Rests on very uneven surface - - - - - ½ to | 2½ |
| | d. Yellow, orange, and red current-bedded sand, with seams and a few pebbles or "galls" of grey clay: a discontinuous layer of fine-grained flaggy ironstone (2 to 4 ins. thick) near the middle in sections (2) and (3)- to | 6 |
| | e. More evenly stratified grey and red sands and loams, with bands of greenish-grey fire-clay, and some soft crimson ironstone or "ruddle." This bed thickens to N.W., and varies much in character - - - seen | 9 |
| | f. Red, blue, and orange mottled plastic clay; top gently waved - - - - - - | 10? |

[1] p. 42.

The most interesting of these beds is the conglomerate, or clay-gravel band (*c*), shown in four of the seven sections given in Fig. 7, and visible in other smaller exposures on the south-western side of the hill. The closely-set fragments of soft clay are of many tints—brown, bright red, dull green, and slate blue—and on fresh damp surfaces have the appearance of a richly-coloured mosaic. The bed is evidently due to the destruction of stiff plastic and laminated sandy clays, and to the redistribution of the fragments by currents. Angular chips of flint and a few very small flint-pebbles occur in the sandy matrix.

The compact ironstone in bed (*d*) contains, in section (3), a few almost completely decomposed flints, having the appearance and softness of kaolin, or fine chalk. It may be remarked that flints associated with the more compact sorts of Tertiary ironstone are frequently rotted in this way. Good examples are to be seen in the Pliocene Beds on the North Downs, and in the Reading Series at Studland in Dorset.

The beds lettered (*b*), (*c*), (*d*), and (*e*) in the above account seem to lie in a hollow in the mottled clay (*f*), which probably rests on an inclined surface of older beds. The London Clay in sections (2) to (7) belongs to the larger mass of the formation which forms the upper part of Windmill Hill and follows its northern slope with the local, north or north-westward, dip.

At the abandoned brickyard, about 250 yards north-west of the mill, a fault—indicated in section (1)—with a down-throw of about 7 ft. to the north-west, brings in the smaller mass of the formation, which is shown on Sheet 13 of the old series of the Geological Survey Map, but not on Sheet 254 of the new. This small dislocation was traced by Mr. Whitaker[2] to the north-eastern boundary of the Nettlebed outlier, but there are now no exposures on the line he indicates, nor any signs of the little inlier of Chalk which he observed on the down-throw side of the fracture.

*Nuffield.*

Around Nuffield Church, at the edge of the Chalk escarpment, there is a thin spread of brown sand and loam. No sections are now to be seen, but in 1858, or thereabouts, Mr. Whitaker noticed near the church a shallow roadside pit showing flaggy and rubbly iron-sandstone in red and brown clay and brown sand. Sand for brick-making was once dug here.[3]

*Stoke Row.*

This outlier consists chiefly of mottled clay and brown loams, and the best exposure, which occurs in the brickfield near the north-eastern side, shows about 15 ft. of the former overlain by 5 or 6 ft. of the latter. Here Mr. Whitaker saw also "light-coloured clays and sands, and some white sand, with many flints, large and small, rounded and angular, sometimes in lines, sometimes in patches, chiefly in the clay, and a few small quartz pebbles"[4]—phenomena suggestive of a recurrence of the gravelly conditions obtaining at Lane End.

Mottled clay has been dug in the field to the north-east of the Maharajah's Well; and at Neal's Kiln, south of Stag Hall, a pit shows 12 to 15 ft. of the same rock. A little of the sandy Bottom Bed, with green-coated flints, was exposed in a shallow water-pipe trench, dug in the early part of 1905, at a point about 100 yards south-east of Stag Hall, by the road-side.

The south-eastern part of this outlier seems to be affected by a disturbance somewhat similar to that at Nettlebed, the beds being tilted downwards to the north-west; for the junction with the Chalk, which occurs at 570 ft. above O.D. by Stag Hall, is not more than 540 ft. where the main street of the village crosses the boundary of the outlier, a little more than

[1] According to Mr. J. H. Blake in *Proc. Geol. Assoc.*, vol. xii (1891), p. 206. The fault runs N.E.—S.W.

[2] 'The Geology of Parts of Oxfordshire and Berkshire,' 1861, p. 43.

[3] *Ibid.*, p. 41.

[4] *Ibid.*, p. 42.

a quarter of a mile farther north. The beds beneath the village seem to thicken down the slope towards the combe on the north.

The maximum thickness of the Reading sediments at Stoke Row probably does not exceed 40 ft.

*Greenmoor Hill.*

At the edge of the Chalk escarpment to the south of Woodcote there is an outlier, about 2 miles in length and from one-half to a quarter of a mile in width, the northern half of which lies within the area included in Sheet 254. In this part of the outlier clays predominate, while farther south the beds are more sandy.

Large quantities of mottled clay and some pale grey sand were excavated in making the foundation-trenches for the South Oxfordshire Water and Gas Company's works on the summit of Greenmoor Hill, in 1905-6, but the present writer was too late to see any section there.

In pits (now degraded) on either side of the road running from the southern part of the hill to Woodcote, Mr. J. H. Blake saw good sections of the even junction of the Reading Beds with the Chalk, and at the brick-yard in Claypits Wood, a quarter of a mile east of The Laurels, he noted [1] :—

| | | Ft. |
|---|---|---|
| "Reading Beds | Mottled crimson, grey, and brown clay - - | 16 |
| | White sand - - - - - - - | 5 |
| | Brown and grey clay interstratified with seams of sand (Bottom Bed) - - - - - | 4 |
| Chalk." | | |

In 1893 two sections in the same yard showed a thin band of white pebbles and subangular pieces of flint in a dark grey sandy clay, resting on grey sand and overlain by red mottled clay.

Further south there are many small exposures of mottled clay on the sides of Greenmoor Hill and at other places just within the boundaries of the outlier, but these do not call for particular notice.

*Moulsford Downs.*

A thin and disturbed remnant of Reading Beds, which seems to consist chiefly of yellow sand, caps the highest part of the Lowbury Camp ridge by Warren Farm, 3 miles south-west of Moulsford.

*Park Place.*

In the grounds of Park Place, on the broad spur to the east of Henley, there is a thin spread of brown and light-coloured sand, with a little mottled clay, once worked for bricks. Signs of the green Bottom Bed were seen near the southern end of the outlier. To judge from its position, and from the appearance of the ground, this patch of Reading Beds lies on an uneven, piped surface of the Chalk.

*Warren Row.*

The Eocene Beds mapped at the south-east corner of the Sheet belong to the extensive Wargrave outlier, the greater part of which lies within the area covered by the Reading and Beaconsfield Sheets (268 and 255), to the south and east. Between Cockpole Green and Warren Row the boundary of the Reading Beds is clearly marked by a rise in the ground and by a succession of swallow-holes. The green pebbly sand at the junction with the Chalk can be seen in the sides of the lanes running south from Pudder's Farm and Dean Place Farm.

On the east side of the road through the little village of Crazies Hill, orange and red-brown stratified sand has been dug in a cottage garden. There are small exposures of sand and of mottled clay on either side of the road west of Warren Row ; but the only clear section of the Reading Beds

[1] 'The Geology of the Country Around Reading' (*Mem. Geol. Surv.*), 1903, p. 15.

in this part of the outlier is in a pit opened on the side of a swallow-hole, at the head of the deep indentation of the Reading boundary, at the eastern end of the last-named village. This showed in 1907 :—

| | | Ft. |
|---|---|---|
| | 5. Clay soil. | |
| Drift. | 4. Compact brown and grey sandy clay, with layers and masses of pebbles derived from the gravel on the higher ground in the vicinity – – | 6 to 10 |
| Reading Beds. | 3 Laminated brown sandy clay – – – | 2 to 7 |
| | 2. Large, unworn, green-coated flints and small flint-pebbles, in grey and greenish glauconitic sand cemented in places by calcareous matter – – | 1 |
| | Teeth of *Odontaspis elegans* (Ag.) and *Otodus obliquus ?* Ag., *Ostrea* (fragments). | |
| | 1. Chalk, traversed by tubular cavities or borings filled with greenish sand to a depth of 3 ft. below its even junction with bed above – – – | 10 |

In bed (2) a flint cast of a carinate *Micraster coranguinum* was observed.

The borings in the Chalk vary from ½ in. to 2 in. in diameter. Some descend vertically for a distance of 1 or 2 ft. occasionally throwing out branches directed obliquely downwards; others, of the simple tubular type, describe regular curves; and both are filled with green sand and pebbles from above, or, less often, by plugs of discoloured chalk. The walls of the tubes bear a faint longitudinal striation.

The Chalk here is mined in galleries which run for a considerable distance under the Common to the east. The entrance to these workings is buried, from time to time, by falls of clay and sand from the upper part of the open pit.

At the cross-roads immediately to the east of this pit there is a swallow-hole, which is constantly fed by a little spring issuing from sands near the top of the Reading Beds on the side of Ashley Hill. The water of this spring is strongly calcareous and deposits a firm granular crust of pale brown tufa upon the twigs, pebbles, bits of brick, and other objects lying in its channel. The calcareous matter in solution may be derived from the thin bands of impure limestone which occur at the base of the London Clay. One of these bands may be seen in the hollow lane south of Holly Cross, just within the area of the Reading Map (268).[1]

## London Clay.

The London Clay occurs in only three places in the Henley district, namely, at Lane End, at Nettlebed, and near Warren Row to the south-east of Henley, and it is nowhere well exposed.

It consists of brown clay, and grey-blue clay weathering brown, with argillaceous limestone in nodules, which are generally described as "septaria," though not all possess the true septarian structure. The lowest beds—collectively termed the Basement Bed—are usually brown loams and loamy sands, with thin

[1] The Warren Row quarry and spring are described in *Proc. Geol. Assoc.*, vol. xvii, 1901, pp. 179–180.

seams of stiff clay, a little olive-green glauconitic sand, concretions of clayey limestone, and flint-pebbles. They are frequently fossiliferous, and at their base there is generally a layer of well-rounded flint-pebbles, some of which attain diameters of 6 to 10 inches.

The junction with the Reading Beds is slightly waved, and its disregard for the oblique stratification of those beds shows that some erosion preceded the deposition of the London Clay; but the general parallelism of the basal planes of the two formations over wide areas makes it probable that the erosion in question did not, in this part of the country, involve much more than a planing down of the constructional inequalities in the upper surface of the Reading Beds.

*Lane End.*

Five outliers have been mapped here: two forming the higher parts of the hills to the east and west of the village; the others underlying the hummocky ground on the slopes of Moor End Common to the south. The section of the Basement Bed seen in the sand-pit to the south-east of Lane End Church has already been described (p. 62). In another part of the same excavation, and at a rather higher level, Mr. Whitaker observed (in 1859):—

| | |
|---|---|
| "London Clay. | Stiff brown clay, with *septaria* and broken fossils.<br>Basement Bed. Brown loam with layers of clay; green grains in parts; at the bottom a few flint-pebbles of various sizes. |
| Reading Beds. | Alternations of laminated greenish clay and brown sand, a foot or more thick.<br>Light-coloured sands." [1] |

Blocks of concretionary calcareous mudstone, crowded with shells of *Cardium*, *Pectunculus*, and *Ditrupa*, were to be seen on the floor of the pit in 1899.

Small exposures of the lower part of the formation occur on Moor End Common, but none presents any features of interest.

There is about 50 ft. of London Clay beneath the summits of the wooded hills to the east and west of Lane End, but much less under the Common to the south of that place.

The following list of fossils found in the lowest beds of the formation at Lane End is taken from Mr. W. Whitaker's Memoir on "The Geology of London" [2]:—

Lamna, teeth.
Aporrhais sowerbyi (*Mant.*)
Calyptræa trochiformis *Lam.*
Cassidaria nodosa (*Brand.*)
Fusus sp.
Natica labellata *Lam.*
" sp.
Pyrula smithii *J. de C. Sow.*
Cardium laytoni *Mor.*
" nitens *J. Sow.*
" sp.
? Cyprina morrisii *J. de C. Sow.*
Cytherea obliqua *Desh.*
? Modiola elegans *J. Sow.*
Ostrea pulchra *J. Sow.*
Panopæa intermedia *J. Sow.*
Pectunculus brevirostris *J. de C. Sow*
" plumsteadiensis *J. Sow*
Tellina ?
Lingula tenuis *J. Sow.*
Ditrupa plana (*J. Sow.*)

[1] 'Geology of the London Basin,' p. 207.
[2] 1889, vol. i, pp. 264-265.

*Nettlebed.*

As the mode of occurrence of the London Clay at this place has already been described (p. 70), it will here suffice to notice the principal exposures.

The beds referable to this formation attain their maximum thickness of about 30 ft. under the windmill; and the following section—better than any now visible—was observed by Mr. Whitaker in the slope a little to the west of that building:—

| | | Ft. |
|---|---|---|
| "1. London Clay. | Light brown clay, partly mottled with pale grey, with thin lines of ironstone and a few small flint-pebbles; darker towards the base, and passing into:— | |
| 2. Basement bed [of London Clay]. | *a.* Dull brown clay, occasionally greenish internally, laminated by very thin seams of sand, with a few small flint-pebbles and a little ironstone | 4 or 5 |
| | *b.* Dull grey and brown loam, with layers of chocolate-coloured clay in parts; at the base a thin bed of green sand, with a few small flint-pebbles; water is thrown out from this - about | 3 |
| | *c.* Seam of ochreous clay - - - - } about | 1 |
| | *d.* Brown sand, with thin lines of ironstone } | |
| | *e.* Line of rounded flints, many very large. | |
| 3. Reading Beds." | | |

From the ironstone in the London Clay proper, Mr. Whitaker obtained "a cast of a small bivalve (*Leda* ?)"; and from that in bed (*a*) "a cast of *Natica.*" There were "lying about some pieces of flat sandy laminated limestone, not more than 2 in. thick, which probably belong to the Basement Bed. The junction of the London Clay and the Reading Beds [showed] a dip of about 6° in a direction 20° to 25° S. of E."[1]

Flaggy and dogger-like masses of the laminated limestone above mentioned, some measuring 3 ft. in length and a foot or more in thickness, are still to be seen on the floor of a small working to the west of the mill; and one of the blocks is visible *in situ* in the same pit. The stone is of a light brown colour and very tough. The constituents insoluble, or not readily soluble, in cold dilute hydrochloric acid are, minute black and red-brown granules (apparently of iron oxides), quartz-grains (mostly about 0·1 millimeter in diameter), and yellow mud.

Faint impressions of shells (like *Pectunculus*) occur on the surfaces of a few of the concretions.

In the brickyard extending across the second (and unmapped) patch of London Clay, a quarter of a mile north-west of the windmill and on the down-throw side of the little fault (*see* p. 69), Mr. Whitaker saw:—

| | | Ft. |
|---|---|---|
| "1. | Drifted clay and loam, in parts with many flint-pebbles, filling hollows in | |
| 2. | London Clay - - - - - - about 2 or | 3 |
| 3. | Basement Bed:—Brown loam, with a few seams of clay, and a little ironstone, but no flints or green sand - - about | 8 |
| 4. | Light-coloured sand of the Reading Beds just visible beneath. | |

The dip shown by the upper part of the Basement Bed is from 5° to 15° a little to the west of south."[2]

[1] 'The Geology of Parts of Oxfordshire and Berkshire' (*Mem. Geol. Surv.*) (Sheet 13, Old Series), 1861, pp. 52-53; see also 'The Geology of the London Basin,' 1872, pp. 205 and 305.

[2] *Ibid.*, p. 53; also 'The Geology of the London Basin,' p. 305.

The greater part of the Basement Bed, with the sand below it, was still to be seen there in 1892, but the section has since become much obscured.

### *Warren Row.*

Part of the London Clay outlier of Bowsey Hill is shown in the south-east corner of Sheet 254. There is from 100 to 130 ft. of the clay beneath the summit of the hill. This is probably but little less than the full thickness of the formation for this part of the country, for recent excavation has proved the presence of dull reddish-brown and pale grey sands and sandy clays, closely resembling those of the Lower Bagshot Beds, beneath the Pebble Gravel on the hill-top. Unworn blocks of sarsen-stone, up to 3 ft. in length, have been observed in the gravel there.[1] There are no good exposures of London Clay, either near Warren Row or in any part of the outlier.

1 *Proc. Geol. Assoc.*, vol. xii. (1892), p. 380.

# CHAPTER IX.

## SCENERY AND SUPERFICIAL DEPOSITS.

By H. J. Osborne White.

### Note on the Principal Physical Features.

In post-Eocene times the rocks of this district, in common with those of the surrounding country, underwent a gentle and probably exceedingly slow and intermittent tilting, which gave them their present prevailing south-eastward dip. As a result of this movement a south-eastward drainage was engendered, and a series of sub-parallel streams were marked out on the constructional slope. Of these early consequent streams—some few of which may have formed the extended trunks of older drainage systems developed in a north-western land-area in Eocene times—the transverse section of the Thames between Oxford and Reading, and perhaps also the Wye, whose valley breaches the Chalk escarpment to the north of Saunderton Lee, may be regarded as the descendants, or modern representatives.

With the uplift which carried marine sediments of Eocene age far above sea-level in this part of the country, the south eastward-flowing streams sank their channels into the inclined rocks, and, encountering strata of varying degrees of resistance in their downward progress, developed lateral or subsequent branches (such as the Thame and the Mill Brook) along the strike of the weaker beds; thus initiating the longitudinal vales, and intervening hill-ranges of the *cuesta* type, which characterise the counties of Bucks and Oxon, and the north-western part of Berkshire. The carving of the existing topography out of the sloping block of Tertiary and older strata has been, without doubt, a most intricate process. Epochs of elevation have alternated with epochs of still-stand or depression; and the local relief, while fluctuating in strength with the varying base-level of erosion, has gradually changed in form with the dipward retreat of the escarpments. The thickness of the Tertiary rocks originally covering the Chalk of the Chilterns is unknown, but it probably greatly exceeded that of the Reading Beds and London Clay (about 100 ft.) remaining in the outlier of Lane End. The exposure of the Upper Jurassic clays and limestones about Radley and Chislehampton, and along the course of the Haseley Brook, has involved the removal of about 1,000 ft. of Cretaceous strata alone.

The escarpment of the Middle and Upper divisions of the Chalk has reached that advanced stage of development in which foundering seldom occurs, and recession is carried on almost entirely by superficial wasting and solution. Its character, however, is markedly dissimilar in different parts of its extent within this district. On the south-west, near the Thames, where recession has been greatest and a condition of comparative stability has been attained, the scarp-face is deeply fretted by narrow branching combes, separated by finger-like spurs with undulating crests, and presents an appearance not unlike that of the most dissected portions of the dip-slope. Farther from the river, by Watlington and Shirburn, the branching combes are replaced by amphitheatral embayments, and the intervening spurs are reduced almost to cusps; while the descent from the crest to the foot of the scarp is more rapid. Still farther north-eastward, near Crowell, the slope is very steep, and the salients and re-entrants are so reduced as to leave the scarp only a moderate sinuosity in plan. Here the escarpment has a fresh, cleanly-cut appearance, and is evidently in a less mature condition than in the south-western part of the area. Indications of slips have been noticed on Bald Hill and Aston Hill (pp. 39, 41).

Although the escarpment of the Chilterns must have changed materially in form, its upper limit, at least, seems to have receded very little, during the development of the minor valley-systems of the dip-slope between the Thames and the Wye; for, in spite of the fact that many of the upper branches of the more important of these systems run back to the crest-line of the upland, scarcely any of them give evidence of former extension beyond the present water-parting. There are two or three very shallow "swales" in the summit, to the west and north-west of Stokenchurch, but not a single depression that could be termed a "gap" breaks the even and gently-inclined crest-line of the range in the 14 miles which separate Chinnor Hill from Greenmoor Hill.

From the considerable dimensions of the gap at the head of the Wye Valley, it is to be inferred that that valley continued to carry the drainage of a tract of country lying to the north-west of the general course of the Chiltern watershed (as roughly defined by the heads of the longer dip-slope valleys, which do not breach the escarpment) down to fairly recent times. A similar state of things still obtains to the north-west of Marlborough, where the Kennet rises on a sort of shelf of Lower Chalk extending beyond (*i.e.*, to the west of) the Middle and Upper Chalk escarpment, through which the river passes in its eastward course.

## Pebble Gravel.

Mixed with the clayey soil and drift over parts of the Chalk, and lying in thicker masses on the summits and flanks of the Eocene outliers, there is a gravel consisting, in the main, of flint pebbles, subangular flints (usually bleached), and small, worn pieces of quartz, together with pebbles of dark, compact, siliceous rocks (cherts and lydites), of light-coloured quartzite, and of other materials, as accessories. Though commonly sandy, the matrix of the stones often partakes somewhat of the nature of the underlying rocks or associated drift.

Being of little economic value, the gravel is seldom worked, and most of the exposures occur in the sides of sand- and clay-pits deliberately opened in places where the overburden of drift is small. In such sections it rarely shows any structure, and even in the few pits where it is dug for road-metal or for levelling uneven ground, no bedding other than that attributable to rainwash, creep, or subsidence is apparent.

The texture of the drift is very variable, and the relative proportions of its constituents not less so, but the qualitative changes in composition are less marked, and it is only in the southern and south-western deposits, near the valley of the Thames, that one finds any notable departure from the normal type. There the red quartzites and other materials more characteristic of the Plateau Gravel appear amongst the flint and quartz of the Pebble Gravel; and in the Reading district, to the south, there are deposits which may, individually, be referred to both classes of drift.

The origin of the Pebble Gravel of the Chilterns has been the subject of a long but fitful controversy, which cannot be reviewed in this place. In the opinion of the present writer, the most likely explanation yet offered is that which ascribes the drift to the wasting of Lower Eocene Pebble-beds and quartziferous gravels of the kinds exemplified in the Reading outlier of Lane End.[1] The Reading Beds at and near that place contain all the commoner, and some of the rarer, constituents of the simpler and more widely represented type of Chiltern Pebble Gravel; and although the average size of some of the materials (*e.g.*, the subangular flints) is smaller than in the Pebble Gravel, there is, as will shortly be seen, evidence of the former existence in this district of Eocene gravel quite as coarse as any in that drift.

It is not, however, contended that all, or, indeed, that most, of the deposits of Pebble Gravel have a strictly local origin. The drift occurs on the summits of isolated hills of London Clay, and, on the assumption that the quartz and lydite pebbles in deposits thus situated are derived from the Reading Beds, it is evident that those materials, at least, must have travelled from places where the parent strata attained a greater altitude—presumably from places farther up the dip-slope of the Chiltern Hills.

The writer has elsewhere[2] given reasons for thinking that the gravelly facies of the Reading Beds may have been a more common one in that part of the formation which once extended north-westward of the present escarpment of the Chilterns, and has suggested that those vanished beds were the chief source of the Pebble Gravel in the country to the south-east of that feature.

If the simpler sort of Pebble Gravel found in this part of the country is a product of the subaërial wasting of Eocene rocks, it cannot be assigned to a definite epoch. Its formation may

[1] C. Reid, in *Summary of Progress of the Geol. Survey for* 1899, p. 140, and H. J. O. White, *Proc. Geol. Assoc.*, vol. xix, 1906, p. 376.

[2] *Proc. Geol. Assoc.*, vol. xix, 1906, p. 376.

have commenced in Oligocene times and have continued ever since; and the constituents of existing deposits may have been dispersed and re-grouped again and again during their intermittent progress down the dip-slope of the country on the northern side of the London Basin.

Some of the principal developments or exposures of the Pebble Gravel will now be noticed.

In the wood named Penley Hollies, 1½ miles south-east of Stokenchurch, pits near the high road show 7 or 8 ft. of pebbly gravel mixed, and irregularly interbedded, with red-brown loam. The deposit appears to occupy a hollow (or group of pipes) in the Chalk. East and south of this a similar drift, more or less mixed with Clay-with-flints, overspreads the Chalk on a moderate slope.

The London Clay hills to the east and west of Lane End have a thin capping of the gravel. It was formerly exposed to a depth of about 5 ft. in a saw-pit near the top of Park Hill, to the east of the village.

The spreads of Pebble Gravel mapped to the south-east of Frieth might be described as pebbly Clay-with-flints.

The more northern of the two patches near Ibstone appears to be the débris of an Eocene flint-pebble bed. The other is of the ordinary quartzose type. Both are thin and mixed with the soil. On the Reading Beds and on the Chalk in the long branching spur of Summer Heath, Garrett's Wood, and Fawley, west of the Hambleden Valley, there are many small patches of very shingly flint-gravel with a varying percentage of quartz-pebbles. One of the best exposures is at Lower Wood End Farm, north of Fawley, where a pit (in an unmapped deposit) shows about 6 ft. of compact gravel. Though close to the edge of the Thames Valley the drift consists of little but flint and quartz, small pebbles of the latter rock being very abundant.

In Queen Wood, south-east of Christmas Common, a shallow lane-cutting, rather above the 700 ft. contour, shows small sections of red-brown, pebbly, Clay-with-flints containing rough blocks of conglomeratic iron-sandstone up to 2 ft. in diameter. The stones in the ferruginous cement include:—

Flint-pebbles.
Bleached sub-angular flints, up to 8 in. in diameter.
Quartz-pebbles, mostly less than ½ in.
Concretions of red iron "pan"; some enclosing pebbles of soft brown clay, others empty.

Though of coarser texture, the constituents of the blocks are the same as those of the ferruginous gravels occurring in the Reading Series at Lane End, and the writer has little doubt that these masses of conglomerate are derived from beds of Lower Eocene age. Pebbles of clay are common in the Reading sands, but have not been seen in any of the upland drifts in this district.

To the east of the hamlet of Russell's Water the old brickyard pits expose a few feet of loamy Reading Beds, thinly capped by Pebble Gravel containing many pieces of ironstone, spongeous green-coated flints, bits of silicified *Inoceramus*-shell, and small seams and bodies of very coarse quartz sand and little quartz-pebbles. This drift is almost certainly the little-travelled débris of gravelly Reading Beds of the Lane End type. On the Common, to the south-east, the drift-gravel is much thicker and coarser, and has been worked to a depth of 12 ft. without being bottomed. A pit recently opened or enlarged shows considerable masses of coloured, thinly-bedded, sandy clay, much like that of the Reading Beds, embedded in and piped by coarse gravel consisting mainly of flint-pebbles and white sub-angular flints (up to 8 or 9 in. in diameter), with a few larger angular flints, some blocks of ironstone conglomerate, and a relatively small proportion of quartz-pebbles. The confused state of the drift here may be due to the formation of pipes in the Chalk below, but some of the phenomena are unlike those produced by mere subsidence.

On the hillock named Priest's Hill, to the north of Nettlebed, old furze-grown pits by the road to Park Corner show a little pebbly loam resting on grey clay and loam of the Reading Beds. Thicker deposits, of very limited extent, are to be seen in the abandoned brickyard a quarter of a mile to the north; and pebbles of flint and quartz are scattered through the loamy soil over the London Clay on Windmill Hill.

At Witheridge Hill, nearly 2 miles south of Nettlebed, small spreads of flint- and quartz-pebbles, in a clayey or loamy matrix, occur on the Chalk.

The gravel resting irregularly upon the Reading Beds, south of Stoke Row, is of a more complex type than any so far noticed; for, in addition to flint, quartz, and lydite, it contains a small proportion of red quartzite pebbles and a variety of sandstones, grits, dark veinstone, and other rocks in pebbles similar to those occurring more abundantly in the Plateau Gravel to the south and south-east, but of smaller average dimensions.

A gravel of the same kind occurs to the south of Checkendon, and in thin spreads or small bodies amidst the Clay-with-flints around that place.

A simpler type of Pebble Gravel, containing a greater proportion than usual of white sub-angular flints, overspreads Greenmoor Hill, near Woodcote; while to the south and east, at a lower level, the red quartzites and sandstones again appear in the loamy soil.

The Pebble Gravel on the narrow ridges of Moulsford Downs, to the west of the Thames, also belongs to the more complex drift, which forms a link between the normal quartzose Pebble Gravel of the Chilterns and the Plateau Gravel, with many quartzites, bordering the Thames Valley.

## Clay-with-Flints.

The widespread drift described in the Index to Sheet 254 as "Clay-with-flints and Loam overlying Chalk," is a product of the decomposition of the Chalk and the disintegration of the Eocene and later sediments formerly covering that rock. It consists, as a rule, of brown to rich red-brown slightly sandy clay; stiff and "cold" in the lower parts, more loamy in the upper, and containing angular or broken flints, flint-nodules, and pebbles of flint and of other materials.[1]

The relative proportions of the constituents vary from place to place, and while the clay is generally the dominant element, the sand, which is never wanting, is sometimes present in sufficient quantity to give the drift a loose and loamy character. The flints, also, are occasionally so closely set as to be workable as gravel. Resting, as a rule, upon an uneven and deeply-corroded surface of chalk, the Clay-with-flints is subject to rapid changes in thickness, as the mottled brown and white appearance of newly-ploughed fields on the higher slopes of the Chalk combes often very clearly shows. It is best developed on the summits of the broader ridges, but commonly descends

[1] See A. J. Jukes-Browne, *Quart. Journ. Geol. Soc.*, vol. lxii, 1906, p. 132.

some distance down their side-slopes, passing insensibly into a loamy soil often very full of stones; and some patches occur at the bottom of the valleys.

Beyond an obscure banding, and certain indistinct vertical and lateral changes in the texture or proportions of its constituents, probably due to internal movements, the drift rarely exhibits any structure in sections.

This flinty clay is, doubtless, in process of formation at the present day, for the hills of Eocene strata are continually shedding their clays and sands on to the surrounding Chalk, and the Chalk itself is slowly yielding up its flints and fine brown muddy residue under the attack of rain-water charged with carbonic and other acids derived from the atmosphere and from dead and living organic matter in the soil.

There are some good sections of Clay-with-flints, in pits from 5 to 10 ft. deep, on the edge of the Chalk escarpment at Chinnor Hill. Broken, unworn flints are very abundant, and the stiff clay in which they are embedded is of a dark red-brown colour with black markings. With the angular flints are unworn nodules and pebbles of the same substance, and flaggy bits of ironstone. Here and there pinnacles of rotted chalk, penetrated by veins of clay, rise nearly to the surface of the ground, and the intervening drift has a rudely synclinal structure.

A more loamy drift, with fewer stones, is exposed in the sides of the road-cutting near the top of Aston Hill.

At the north-east corner of Hailey Wood, to the south of Aston Hill, there are shallow pits in sandy loam (looking much like re-arranged Reading Beds) containing the usual flint fragments and pebbles.

In Shotridge Wood, south of Shirburn Lodge, degraded pits show a little red-brown loam with many bleached angular flints and bits of compact ironstone.

Drift of a more argillaceous character caps the irregular spur extending from Watlington Hill nearly to Henley. A few red quartzite-pebbles and bits of sandstone occur in it to the south-west of Fawley.

Loam with few flints has been dug to a depth of 8 ft., or more, on both sides of the road on the east of Watlington Park. A sample of unctuous red clay, taken from the foundation-trenches of a new house about one-third of a mile south-south-west of Christmas Common Church, contained small pieces of decaying *Inoceramus*-shell and minute concretions of iron oxide. The clay itself was exceedingly fine, and may well have been derived from the Chalk. A little fine quartz sand was present.

The banks of the pond at Hollandridge Farm consist of red mottled clay (evidently derived from the Reading Beds), with seams of flinty loam. The occurrence of ironstone conglomerate in the Clay-with-flints about a mile to the north-west of this place is noticed above (p. 80). Blocks of this stone are set up as corner-posts at the entrance to the farm.

Where the road from Howe Hill to Cookley Green crosses the head of the combe running south-eastward to Pishill flinty loam has been dug to a depth of 20 ft.

On Russell's Water Common roadside pits nearly half a mile north-west of Oak Farm showed a mass of angular flints and pieces of ironstone in a brown loam. More argillaceous loam, like that of the Reading Beds, underlies this deposit, which is within the boundary of the Eocene outlier of Russell's Water, as shown on the map. Though containing some pebbles, this drift is quite distir ct from the gravel at the northern end of the Common (see p. 80). Flinty loams are not uncommon on the thinner outliers of the Reading Beds in this and other districts. They occur on the doubtful outlier of those beds to the south-east of Ibstone.

The lower spreads of Clay-with-flints between Bix and Woodcote contain many ill-defined bodies of gravel derived from deposits of the Pebble and Plateau series in their neighbourhood.

The deposits on the summit and north-eastern slope of the blunt spur to the east of Henley present no features of particular interest.

On the chalk spurs of Moulsford Downs, in the south-western corner of the district, there are a few patches of red-brown loam and clay, more or less mixed with Pebble Gravel.

It may be mentioned that the late Mr. J. H. Blake recorded (on the 6in. field-maps of the Geological Survey) the measurements of many of the larger blocks of sarsen and ironstone scattered over the district. Sarsens over 4 ft. in length are very rare on the Chalk uplands of the Henley area, though blocks of far greater size abound in the drift loams of the Beaconsfield district, immediately to the east.

# CHAPTER X.

## SUPERFICIAL DEPOSITS (*continued*).

BY H. J. OSBORNE WHITE.

### PLATEAU GRAVEL.

The drifts which have been mapped as Plateau Gravel form two distinct groups, the members of the one differing from those of the other in their composition and in their relations to the larger physiographical features. One of these groups comprises the angular flint-gravels which range along the lower slopes of the Chalk escarpment from Ipsden, northward, by Ewelme to Brightwell Uppertown, and cap the Upper Greensand between Berrick Salome and Clare Hill. The other includes the high-level, slope and terrace gravels which contain numerous pebbles or cobbles of quartzite and sandstone, and which occur on both sides of the Thames Valley in the south-eastern part of the district near Henley; together with some smaller isolated patches that lie, for the most part, within a mile and a-half of the Thames between Nuneham Courtenay and Cleeve, on the west. As the latter group contains many deposits which seem to be considerably older than any in the former, it will be dealt with first.

#### 1. *The Quartzite Gravel.*

The constituents of this drift include (more or less in the order of abundance) flints, in all conditions, but mostly sub-angular and blanched or stained brown; ovoid and plano-convex cobbles of red, liver-coloured, brown, and grey compact quartzite; sub-angular and rounded pieces of brown, yellow, and dull red sandstone and grit, often felspathic; blocks of sarsen and of iron-sandstone; white vein-quartz, in pebbles and cuboid blocks; together with rolled pieces of jasper, veinstone, lydite, dark chert, mudstone, quartz conglomerate, tourmaline grit, and occasional fragments of devitrified rhyolite, and of many other siliceous rocks. To the east of Reading, flaggy pieces of Lower Greensand chert and sand-stone, like those which occur so plentifully in the drifts of the Loddon basin, make their appearance, and in some places form no small proportion of the coarser constituents of the gravel. The matrix of the stones varies in character from coarse loose sand to stiff clay.

Plateau Gravel of this type has a wide distribution in the lower or eastern parts of the Thames basin, but becomes restricted westward to a belt of country, a few miles in breadth, bordering and including the valley of the main stream between Maidenhead and Oxford.

The Bunter Beds of the Trias, exposed in the country to which these passes give easiest access from the south-east, contain an abundance of quartzite and other rock pebbles identical in character with those in the gravel here under notice, and have long been regarded as the source of the distinctive constituents of that drift.[1] At the present time, the Cotteswolds form, in a general way, the water-parting between the Thames and the Severn, and the nearest outcrop of the Bunter Beds lies well within the basin of the latter river, but recent studies of the geomorphological features of those hills, and of the country about them, have elicited a number of facts which strongly support the view[2], that the Thames formerly drained a considerable area to the north-west of the present divide: hence it has been inferred that the high-level quartzite-bearing gravels of the Upper Thames basin—exhibiting, as they frequently do, both the structure and disposition of river gravels—were formed by streams which derived a part of their water from Triassic terrains in the Trans-Cotteswold area above mentioned.[3]

It is not improbable that the transportation of the Bunter pebbles across the line of the existing Severn-Thames divide was accomplished in part by ice, as many geologists have supposed, for deposits which seem to be modifications of a common type of Midland boulder-clay occur on the floor of the Moreton gap, at the head of the Evenlode Valley.[4] That the quartzitic Plateau Gravel bordering the Thames Valley is largely of fluviatile origin can, however, hardly be doubted; and, although proof is lacking, the present writer is much disposed to think that the exotic material which occurs so abundantly in that drift, and which, in the Henley district, appears at first in rounded stones of small size in the upland Pebble Gravel, 400 to 500 ft. above the Thames, was being carried into the region of the Upper Thames basin long before the commencement of the Pleistocene 'Ice Age.'[5]

1 See W. Buckland, '*Reliquiæ Diluvianæ*' (1821, 4to), p. 249, and W. C. Lucy, 'On the Gravels of the Severn, Avon, and Evenlode, and their Extension over the Cotteswold Hills.' *Proc. Cottesw. Nat. Club*, vol. v, 1869, p. 71.

2 First expressed by Dr. T. S. Ellis in a paper 'On Some Features in the Formation of the Severn Valley as seen near Gloucester.' *Gloucester Phil. Soc.*, 1882. See also, W. M. Davis on 'The Development of Certain English Rivers.' *Geogr. Journ.*, vol. v, No. 2 (Feb., 1895), pp. 127-146; and on 'The Drainage of Cuestas.' *Proc. Geol. Assoc.*, vol. xvi (1899), p. 75. S. S. Buckman, 'The Development of Rivers,' &c. *Nat. Sci.*, vol. xiv (1899), p. 273.

3 H. J. O. White, On the origin of the High Level Gravel with Triassic débris, &c. *Proc. Geol. Assoc.*, vol. xv, 1897, p. 157. It should have been stated that in the sketch-map illustrating this paper, the presence of Triassic material near Cirencester was indicated on the authority of Prof. E. Hull and Mr. W. C. Lucy, whose observations are, however, noticed in the text (p. 161).

4 H. B. Woodward, 'The Chalky Boulder Clay,' &c. *Geol. Mag.*, Dec. iv, vol iv (1897), p. 485; and T. Mallard Reade, S. S. Buckman, and C. Callaway, 'The Gravel at Moreton-in-the-Marsh,' *Proc. Cottesw. Nat. Field Club*, vol. xiv (1903), p. 111.

5 Glacial, and extra-glacial lacustrine, origins have lately been claimed for some of the high-level gravels around Oxford. See works by Mr. T. I. Pocock and Mr. F. W. Harmer mentioned in the bibliographical appendix to this Memoir.

In the Henley district, as elsewhere, the plateau, or high-level gravel, with many quartzites, occurs both in slightly-inclined flats or terraces, and in more strongly sloping sheets, which may, in some cases, have resulted from the degradation of terrace deposits. In the more nearly level spreads the gravel usually has a sandy matrix and is occasionally stratified, but in those with a pronounced slope the stones are more often embedded in a loam or clay, which is either devoid of structure, or shows only an impersistent banding, such as might arise from creeping movements.

The gravel is decalcified and, save for a few palæoliths, has yielded no fossils. As to its age, little is certainly known. The lower terrace deposits—*i.e.*, those up to 200 ft. above the level of the Thames in their neighbourhood—are probably all of Pleistocene date; the highest may be of Newer Pliocene age.

Where it traverses the broad depression marking the basset surfaces of the Selbornian Rocks and the Kimeridge Clay, in the western part of the district, the valley of the Thames is shallow and ill-defined, and most of the coarse drifts along its course are low-lying deposits, which nave been mapped as Valley Gravel. Some small bodies of higher gravel, however, occur on the hillocks of Lower Greensand and Gault in and around Nuneham Park, and to the north-west of Chislehampton.

The Sinodun Hills, which rise sharply from the river bank south of Dorchester, and form such conspicuous objects in the midst of the plain at the meeting of the Vales of Whitehorse and Aylesbury, also bear two patches of the drift. One of these (not shown on the map) covers the summit of Harp Hill (Wittenham Clump), the highest point of the little range, about 250 ft. above river-level, and a larger spread occurs on the eastern slopes of Brightwell Barrow.

Farther south, near Moulsford and Cleeve, at the entrance to the so-called "gorge" of Goring, thin deposits of quartzite gravel occur on the ends of the spurs of the Chalk escarpment, on either side of the river. The largest of these, to the south-east of Cleeve, lies on a distinct terrace, about 220 ft. above the river.

In the Reading area the Plateau Gravel spreads over much of the higher ground below the 500 ft. contour-line on the northern, and, to a less extent, on the southern, side of the Thames Valley, which there describes a bold curve through east to north. The upper limit of the gravel-sheets to the north of the river enters the Henley district near Exlade Street, and runs thence north-eastward, athwart the general slope of the Chilterns, to Bix, beyond which place it trends eastward, by Fawley and Hambleden. In many places along the line indicated there is an appreciable change in the form of the ground between the branching combes; the steadily-sloping country occupied by the larger spreads of Clay-with-flints and the patches of Pebble Gravel, on the north and north-west, being separated from the more gently-sloping tract occupied by the principal sheets of Plateau Gravel, on the south and south-east, by a belt of ground possessing a stronger inclination. Thin and degraded-looking deposits of the gravel generally occur within this belt, and may even extend across it; but in a few places, where the slope is more pronounced than usual, the drift stops short at its lower limit; which limit, it should be remarked, varies considerably in altitude above both sea- and river-level.

Most of the deposits above 270 ft. O.D. to the south-west of the Assendon Valley (which joins the Thames Valley at Henley) lie on rather strong slopes of 60 to over 100 ft. per mile, and many of them are very argillaceous and mixed with Clay-with-flints. There are few good exposures of these higher gravels. Small pits occur in Lackmore Wood, south of Exlade Street, and there are many little workings round the borders of Kingwood Common, above 370 ft.

At Shiplake Hill (340 ft.) a pit on the north-east side of the road to Peppard Common formerly showed about 15 ft. of gravel with seams of red-brown sand and loam. Near the top the stones were small and mixed with the loam; lower down the drift was coarser and distinctly bedded.

A roadside pit at Shepherd's Green, a little above 400 ft. O.D., exposed (in 1894) about 9 ft. of red-brown sand with seams of quartzite gravel. The same kind of drift was seen in a pit about 8 ft. deep in the wood to the south-east of Highmore Cross. There, part of a lens of coarse gravel was just visible at the bottom of the section. In the same wood a shallow cutting for a drive exposed a few feet of coloured clay and clayey gravel, with what seemed to be a flow-structure.

At Bix, near the church, the gravel covers a little flat or terrace. Near the upper limit of the drift to the north-west of that place the pebbles of exotic rocks are small and well rounded.

The lower deposits to the south-west of the Assendon Valley, and many of the higher deposits to the east of that feature, occupy distinct terraces at various heights above Ordnance Datum and river-level.

*Fig. 8.—Diagrammatic View and Section of the Thames Valley North of Henley-on-Thames.*[1]

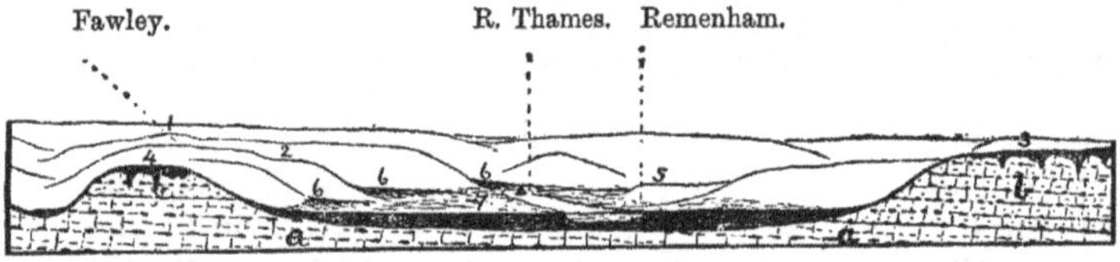


"*a.* Middle Chalk.
*b.* Upper Chalk.
1. Pebble Gravel on Reading outlier, 520 ft. O.D.
2. Terrace Gravel. Benham's, 400 ft.
3. Terrace Gravel. White Hill, 340 ft.
4. Terrace Gravel. No Man's Hill, 315 ft.
5. Terrace Gravel. Remenham, 180 ft.
6. Rubble Drift Fans.
7. Valley Gravel and Alluvium."

NOTE.—On Sheet 254 the gravels 2, 3, and 4 are mapped as Plateau Gravel; and 5, 6, and 7 as Valley Gravel. The Thames is here about 100 ft. above O.D.

The highest of these terraces is that at Benham's, to the south-east of Fawley. Its surface stands a little above the 400 ft. contour near the

[1] H. J. O. White, *Proc. Geol. Assoc.*, vol. xviii (1904), *p.* 415, fig. 21.

cottages shown on the map, and is there almost level. To the north, the ground rises quickly to the 500 ft. contour by Fawley Green, and to the south and south-east there is a steeper fall, of 300 ft., to the Thames. The gravel has been worked, to a depth of 10 ft., in a pit at the southern margin of the terrace.

Remains of another terrace, about 30 ft. lower than the last, occur on the ends of the spurs north-east and north-west of Greenlands, to the south-east of Dobson's Farm, and at Great Hill in Henley Park. The gravels above Greenlands are pebbly.

Still lower, at about 330 to 340 ft. above sea-level (and 100 ft. less above the river), there is a dissected terrace about Bockmer and Burrow Farm, to the east of Hambleden. The flat is bounded on the north by a distinct rise in the ground, up which the drift extends for some distance; and on the south by the steep-sided trench of the Thames Valley. A small pit, south-west of Rotten Row, on the sloping ground to the north of the terrace-feature, showed 2 or 3 ft. of red-brown sandy gravel, in which pieces of quartz and quartzite appeared to be as numerous as those of flint.

The degraded deposits, below 350 ft. O.D., at and to the east of Remenham Place, on the south side of the river, may belong to the same stage as the Bockmer terrace. Bits of Lower Greensand chert and cherty sandstone are common in the drift which fills the pipes exposed in the chalk-cutting on White Hill, east of Remenham Hill.

On the spur called No Man's Hill, which overlooks the town of Henley from the north, there is a flat, at 315 to 320 ft., which is dominated by the higher ground of Great Hill.

From a pit in Friar Park, at the lower end of the sheet of gravel which slopes eastward from 350 ft. near New Farm to about 230 ft. at Henley, Mr. Ll. Treacher, of Twyford, has obtained (through the workmen) a well-worked palæolith of the "pear-shaped" type. Other implements are said to have been found at the same spot.

South-west of Henley, a clearly-defined terrace between 255 and 260 ft. above sea-level (160 ft. above the Thames) supports Gillott's House, Coldharbour Farm, and Herne's Farm. A few palæoliths have been found near Highlands Farm..[1] The outer edge of this flat is marked by a low bluff, where the strip of bare chalk is shown on the map. The slope below the terrace, to the west of Henley Newtown, is in places thickly covered with gravel and gravelly loam, as the numerous excavations which have been made in it since the district was geologically surveyed have demonstrated.

In the small patch on sloping ground, between 250 and 225 ft., around the site of the Castle north of Medmenham, there are pits showing from 6 to 12 ft. of sandy, ferruginous, stratified gravel, containing many pieces of Lower Greensand chert.

From the foregoing remarks it will be gathered that there are five well-marked terraces, or groups of terraces, of the Plateau Gravel, between 150 and 300 ft. above the Thames, near Henley, as well as many sloping spreads with a greater aggregate vertical range. These terraces are all of the same, "cut-and-built," type—platforms eroded in the Chalk and covered by the drift—and there can be little doubt that they are the work of the Thames, and that they mark pauses in the negative or downward progress of the local base-level of erosion. The 160-ft. terrace to the south-west of Henley has its fellows in the Reading district, and is probably contemporaneous with the extensive gravel plateaux (of the present writer's "Silchester Stage"[2]) which occur at about the same elevation above the principal waterways in the Kennet and Loddon basins.

1 *Proc. Geol. Assoc.*, vol. xiv (1895), p. 20.

2 'On a Peculiarity in the Course of Certain Streams in the London and Hampshire Basins.' *Proc. Geol. Assoc.*, vol. xvii (1902), p. 404.

## 2. *The Angular Flint Gravel.*

The following description of the deposits on the lower slopes of the Chalk escarpment, to the north and south of Ewelme, is mainly contributed by A. J. Jukes-Browne.

The Plateau Gravels which occur in the country to the east of the present courses of the Thame and Thames exhibit peculiar features, for they consist almost entirely of flints which have clearly been derived from the Chalk escarpment; and they

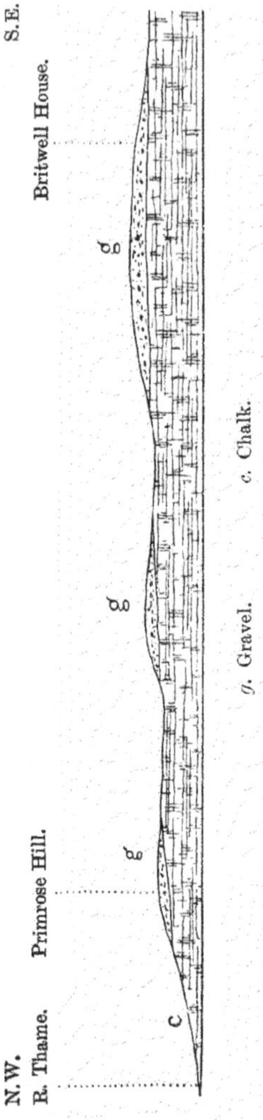


Fig. 9.—*Section from Brightwell Baldwin to the River Thame near Drayton St. Leonard.*

include patches of brown clay which may well have been derived from the Eocene, or from its *remanié* form of Clay-with-flints.

Moreover, they lie on a gently inclined surface which slopes from east to west, and extends over a considerable width of country. Thus, between Britwell Salome and the valley of the Thame opposite Drayton St. Leonard, the patches of Plateau Gravel occur over a tract which is more than 4 miles wide, and on a surface which slopes gradually from a height of about 430 ft. at Brightwell House to 240 ft. on Primrose Hill[1] (*see* section, Fig. 9).

It is possible that these are remnants of successive terraces of gravel belonging to an ancient series of valley gravels, but if so the continuation of the lower terrace has been destroyed in the district east and south of Wallingford, where there are no intermediate gravels between the Plateau Series at and above 300 ft., and the gravels of the modern Valley below 200 ft.

Of the patches at higher levels the most northern is a small patch resting on the Upper Greensand of Clare Hill, two miles north-west of Watlington. No exposure was seen, but the stones on the fields seemed to be all flints. The surface height is 393 ft.

There is another small tract on the hill north of Cuxham of similar gravel at a height of 368 ft.

West of Brightwell Baldwin is another larger tract extending over the fields between Brightwell Park and Whitehouse Farm, its base near this farm going a little below 300 ft. On the eastern half of this tract the soil is clayey, as if a clay overlies sand and gravel; the clay varies in character, being in one place whitish marly clay full of flints and resembling boulder-clay but at the west end of the plantation west of the road to Cadwell it is a sticky yellowish-brown clay. In the plantation is a saw-pit showing a very sandy gravel. The western part by Whitehouse Farm is all gravelly, and a pit by the farm shows that the material consists entirely of flints.

South of Lonesome Farm on the same ridge there is again gravel of the same kind in an elongate patch, and another on the next ridge to the southward. The surface level of the base of the gravel here is about 280 ft.

South of Brightwell Baldwin and west of Britwell there is a large tract of this Plateau Gravel, the western border of which is truncated at a level of about 330 ft., while its south-eastern border near Britwell House rises to nearly 440 ft. This tract covers about two square miles of ground lying chiefly on lower chalk, but passing on to Middle Chalk near Britwell. Gravel has been dug at several places on this plateau, but there was no open pit at the time of my visit in 1885. I was informed, however, that it includes beds of dark brown clay, and that the bricks used in the construction of Britwell House and of West End Villa were made from clay dug near Brockholes Covert, a little to the south-west of the former place; that similar clay occurs in the field in front of Britwell House, and also by the clump of trees opposite the turning into the road to Brightwell Baldwin.

Sir J. Prestwich states[2] that a tooth of mammoth was found in the gravel "near Britwell House."

A valley cut through chalk separates the plateau above mentioned from a small patch extending eastward from Ewelme for about a mile, and lying on a slope which passes in that distance from 300 to a little over 400 ft. south-west of Ewelme Church there were pits showing from 10 to 15 ft. of flint gravel, and another on the road to Britwell in the same material.

South of Ewelme, on Rabbits Hill, another large tract of gravel commences, and its base rises south-eastward till at Potters Farm it is about 430 ft. It is noteworthy that at this place the gravel is banked up against a slope of solid chalk, and we may consequently assume that its boundary is here very near to the original boundary of the deposit, for in other places its border generally rises above the level of the adjacent chalk

[1] The drift on Primrose Hill is mapped as Valley Gravel.

[2] 'The Raised Beaches and Head or Rubble Drift,' etc. *Quart. Journ. Geol. Soc.*, vol. xlviii, 1892, p. 314.

land, and has evidently been cut back by sub-aerial erosion to a greater or less distance from its original boundary.

Mr. White states that a large excavation a quarter of a mile north-west of Gould's Grove shows upwards of 15 ft. of angular flint gravel with seams of sandy loam towards the base.

Southward this tract of gravel extends for a distance of more than two miles, with an average width of three-quarters of a mile. Such exposures as were seen showed gravel composed entirely of flints. The best of these was in a pit about half a mile west of Turner's Court Farm. This was 10 ft. deep in coarse flint gravel, the flints being partly angular and broken, partly weathered, but none of them rolled. A. J. J.-B.

In its lower part the gravel at Turner's Court has a matrix of fine chalk-rubble, usually soft, but occasionally firm and compact. The looser and more sandy drift above is piped into, and contains detached masses of, the calcareous gravel. The section has been described by Mr. O. A. Shrubsole, from whose paper[1] the accompanying figure is taken.

Fig. 10.—*Section at Turner's Court Gravel-pit, Wallingford.* (O. A. Shrubsole.)

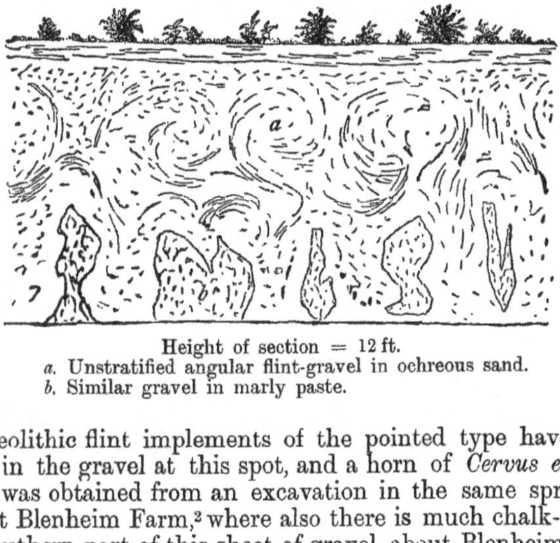


Height of section = 12 ft.
*a.* Unstratified angular flint-gravel in ochreous sand.
*b.* Similar gravel in marly paste.

Palæolithic flint implements of the pointed type have been found in the gravel at this spot, and a horn of *Cervus elaphus* Linn., was obtained from an excavation in the same spread of drift at Blenheim Farm,[2] where also there is much chalk-rubble. The southern part of this sheet of gravel, about Blenheim Farm and Foxberry Wood, lies in a depression bounded by the escarpment of the Upper Chalk on the east and by low hills of Middle Chalk on the west. This depression deepens rather quickly southward, and enters the Thames valley, as a wide-mouthed combe, near Littlestoke House. H. J. O. W.

Half-a-mile south of Turner's Court Farm the gravel plateau ends in a bold ridge which points up the valley toward Nuffield; its abrupt ending and the difference between the gravelly soil and the chalk slopes below being very apparent. The position

[1] 'On Some High-level Gravels in Berkshire and Oxfordshire.' *Quart. Journ. Geol. Soc.*, vol. liv (1898), pp. 595-6, fig. 2.

[2] Prestwich, *loc. cit.*

of the gravel at this place suggests the inference that it originally extended eastward up the valley, and that its materials had been brought down this and other valleys from the summit of the Chalk escarpment. Subsequent valley-erosion has removed its higher portions, and has cut away the continuation of the plateau on which it was spread out, the bottom of the modern valley at Woodhouse Farm being probably 40 ft. below the surface of the gravel to the north of it.

The next tract is found at Hailey, near Ipsden, but is a much smaller one. In this, north of Hailey Farm, a small pit was open in 1885 which showed about 8 ft. of flint gravel. I was informed that in the churchyard there was thin gravel with reddish clay below resting on chalk. North-east of the vicarage is a pit which showed 10 ft. of coarse flint gravel, with some layers of brown loam or clay. The base of this is close on the contour of 300 ft., as is also another small outlying patch by Ipsden House.

Beyond this, toward the gorge of the Thames, few remnants of these old gravels have survived the subsequent erosion, the only one worthy of note being a patch of gravelly soil on Catsbrain Hill, north-east of Grove Farm, at a height of about 250 ft. A. J. J.-B.

The flint gravels above described are, almost certainly, the work of streams, or rather of torrents, which flowed down the combes trenching the upper parts of the Chalk escarpment to the east and south-east. They rest on an uneven surface, and appear to have been spread out in confluent fans extending from the foot of the steepest part of the scarp.

These deposits are not easily to be correlated with other drifts in the district, for the position of the lower limit of the slope or slopes on which they were laid down is unknown. The presence of chalk-rubble among the flints near Turner's Court Farm and at Blenheim Farm; the occurrence of mammalian remains at the latter place and near Britwell House, and the fact that the lowest patches (viz., those of Primrose Hill, near Drayton, and Catsbrain Hill, near South Stoke) are less than 120 ft. above the present level of the Thame and Thames, however, render it probable that they are, on the whole, of more recent origin than the deposits of quartzite-bearing Plateau Gravel described in the earlier part of this chapter. H. J. O. W.

# CHAPTER XI.

## SUPERFICIAL DEPOSITS (*continued*).

BY H. J. OSBORNE WHITE.

### VALLEY GRAVEL.

The deposits mapped as Valley Gravel on Sheet 254 are, for the most part, merely the lower members of a drift-series which includes the gravels dealt with in the last chapter. They cover the floors and portions of the sides of the Thames and Thame Valleys, and of some of the combes and dales opening into those valleys.

The higher bodies of Valley Gravel usually resemble, in all respects, the deposits of Plateau Gravel in their vicinity: the lower are frequently distinguished by a larger proportion of the débris of local rocks, or of rocks which are exposed in the adjoining higher portions of the Thames basin—little-worn flints, pieces of chalk, malmstone, and oolite, and the stouter fossils of the Oxford Clay and Coral Rag occurring more or less freely, according to the locality. Chalk is the chief constituent of some of the low-level deposits in the south-eastern part of the district.

Mammalian and molluscan remains have been remarked in a few places, but Palæolithic tools, which abound in the Valley Gravels near Reading, seem scarce in those about Wallingford and Henley.

Some portions of the Valley Gravel shown on the map appear to belong to the Recent, or Holocene, Alluvium, but the bulk of it is of Pleistocene age.

#### *West of the Chiltern Hills.*

In the western part of the Henley district Valley Gravel covers a considerable area. It occurs mainly in the form of gently-inclined sheets, which rise from beneath the alluvium bordering the present courses of the Thames and Thame to heights of 20, 40, or more feet above those streams. Definite terraces are rare, but, as in the adjoining district covered by the special Oxford map of the Geological Survey, a tendency to form a flat between 30 and 40 ft. above local stream-level is apparent in many places. The higher deposits will be noticed first.

Remains of a terrace about 40 ft. above the Thames occur to the west of Clifton Hampden. A shallow pit, a little more than a quarter of a mile to the south of Culham Farm, shows a few feet of rather fine, well-stratified gravel; containing many worn fragments of *Ostrea, Gryphœa,* and of other fossils derived from the Jurassic strata farther west, and resting on coarse brown sand of the Lower Greensand. Some of the stones in the drift are encrusted with tufa.

North-east of Burcot there are sloping sheets of fine reddish gravel, with much coarse grit from the Lower Greensand, between 60 and 30 ft. above the Thames and the Thame. The lower edge of the principal spread, north and east of Field Farm, is marked by a slight feature where the strip of bare Gault is mapped.

A sheet of coarser flint gravel follows the right bank of the Cuxham Brook from Chalgrove Field, north-westward, to Stadhampton. It forms slightly undulating ground, mostly less than 20 ft. above the stream. The higher spread of Primrose Hill and Starveall, to the south of this, was noticed in the last chapter.

North of Benson, a broad and low ridge or swell of angular and subangular flint gravel runs westward from the foot of Scald Hill towards Hale Farm. Its lower end, near the latter place, is about 40 ft. above the Thames at Benson. This bank of drift seems to be the remains of a delta-like accumulation formed at the mouths of the chalk combes which head near Swyncombe. The gravel is worked near the cross-roads half a mile south-east of Roke.

At Clapcot Fields, north and north-west of Wallingford, a thin sheet of gravel covers a well-marked flat, 30 to 40 ft. above the river.

Spreads of thicker gravel, terraced in places, and rising 20 to 30 ft. above the alluvium, occur on the Chalk and Upper Greensand islands of Mackney and Cholsey Hill, in the marsh-land to the west and south-west of Wallingford.

The Berkshire County Lunatic Asylum, south-east of Cholsey, stands on a gravel-flat about 30 ft. above the Thames. Professor E. W. Claypole noted the following section when the ground was being dug for the foundations of that building:—

| | | Ft. |
|---|---|---|
| "Surface soil | - - - - - - - - | 5 |
| Yellow sand full of comminuted shell [s] too small to be recognised, but seeming to belong to Paludina, Cyclas, and perhaps Lymnæa | - - | 4 |
| Gravel | - - - - - - exposed | 3" |

Professor Claypole also states that two fragments of bone, "one consisting apparently of the end of a tarsus of a species of Bos," were thrown out of a trial pit close by; and that a "perfect antler of the red deer . . . of a very recent appearance" was found on the surface of the Chalk beneath the gravel in the same excavation.[1]

A well made at the Asylum in 1883 proved 13 ft. of gravel over the Chalk Marl[2]; and a slightly smaller thickness of the drift is shown in a pit by the cross-roads to the south-west.

Between the Great Western Railway and the foot of White Hill, at South Stoke, there is a gently sloping terrace at about the same height as the last; and another, with an average elevation slightly greater, is intersected by the railway south of Cleeve.

[1] 'On Some Gravels in the Valley of the Thames in Berkshire.' *Proc. Bristol Nat. Soc.* for 1871, New Series, vol. vi (1872), pp. 30-31.

[2] J. H. Blake and W. Whitaker, 'The Water Supply of Berkshire' (*Mem. Geol. Surv.*), 1902, p. 57.

A boring at Shillingford Brewery, about one-third of a mile from, and 12 to 13 ft. above, the river passed through 15 ft. of gravel, over Gault. (See p. 4.)

Gravel has been dug in shallow pits at Littleworth, near Benson; and Mr. A. J. Jukes-Browne noted a section showing 12 ft. of it over 2 ft. of sand north-west of the smithy at Benson.

A little chalky gravel was exposed in the road bank by the north-eastern corner of Howberry Park, south of Preston Crowmarsh.

At Wallingford, 2½ ft. of gravelly soil was found above the Upper Greensand in the Waterworks well; and 11 ft. of gravel in the well at the Anchor Brewery, farther down the slope to the river.[1]

Tongues of gravel run up the broad combes which enter the main valley from the east near North and South Stoke. A pit half a mile east of Littlestoke House, at the mouth of the more northern combe, shows a good section, 10 ft. deep, of stratified, coarse and fine gravel, made up almost entirely of little-worn, bleached flints, and pebbles and sub-angular pieces of chalk, in a sharp white sand or grit. Bits of *Inoceramus*-shell are common among the smaller materials. The upper layers of the gravel, for depths of 6 in. to 2 ft., are loamy and decalcified. This is obviously a drift of local origin, formed under much the same conditions as (though at a later date than) the flinty Plateau Gravel of Blenheim Farm, in the higher part of the same combe. Near its lower and broader end, the deposit has the form of a very depressed cone, and in crossing it the higher of the two roads connecting North and South Stoke rises faintly from the sides towards the axis of median line of the combe from which the deposit spreads.

Near Shirburn, in the Thame basin, a sheet of angular flint gravel covers much of the south-eastern side of the short strike-valley in which the head-waters of the main branch of the Cuxham Brook are situated.

### *Near Henley.*

In the south-eastern part of the district there are only two distinct terraces of Valley Gravel. One of these—a particularly good example—occurs on the end of the spur in the bend of the Thames Valley at Remenham, and to the south of the Remenham-Aston Lane. Its surface, which stands about 85 to 90 ft. above the river,[2] is almost level, and its inner and outer edges are sharply defined—the former by a steep slope to the floor of the valley, the latter by the rising ground of Remenham Place. A roadside pit, a little to the south-east of Remenham Church gives a good section (10 ft. deep) of well-bedded gravel—of fine to medium texture above and very coarse below—containing many small blocks of sarsen and vein-quartz, and cobbles of quartzite and sandstone. The bedding is disturbed in places by the settlement of the gravel into pipes, such as may be seen in the adjacent chalk-pit in the scarp below the terrace.

The other terrace referred to occurs at the southern end of

[1] 'Water Supply of Berkshire' (*Mem. Geol. Surv.*), 1902, p. 88.

[2] The writer slightly under-estimated the height of this terrace in *Proc. Geol. Assoc.*, vol. xvi (1899), p. 253.

Henley Newtown. It is about 20 ft. above river-level on the line of the high-road to Reading, and the gravel has been dug to a depth of 8 ft., or more, at the turning to the Paper mills.

A pit in the sloping sheet of drift (mostly unmapped) to the west of Henley Newtown, and at about 100 ft. above the river, shows 9 or 10 ft. of red, sandy, stratified gravel. In the side of the valley here there is a slight convexity which probably marks the remnant of a terrace.

The most interesting feature of the lower sheets of gravel in the main valley in this neighbourhood are the cones or fans of chalk-rubble and rubbly gravel which occur at the mouths of the combes to the north and north-east of Henley.[1] These

*Fig.* 11.—*Fan of Chalky Gravel at Greenlands, S.W. of Hambleden.*

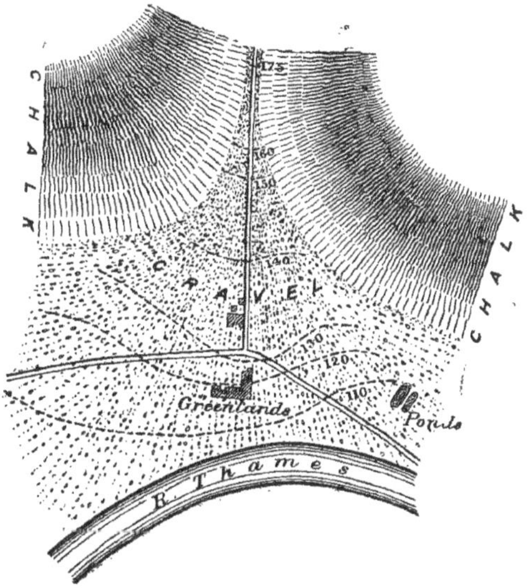


NOTE.—The 10 ft. contours are based on the altitudes given in the 6-in. Ordnance Map. Scale about 4 in. to 1 mile.

are well seen from the Henley-Medmenham road, which gently rises and falls as it passes across them and the intervening depressions. The fans spreading from the shorter combes are steeper and more easily recognised than those from the longer; and the clearest example is that which is radially bisected in plan by the farm road running north from Greenlands (Fig. 11). Near Hambleden Lock, the very gently-inclined fan from the Hambleden valley is scarped by the Thames, and a portion of its transverse profile is well shown in the upper edge of the low bluff which runs north-westward from the mill. Much of the eastern half of this spread has been worn away by the Hamble-

[1] See *Proc. Geol. Assoc.*, vol. xviii (1904), pp. 417-418.

den Brook. The limit of the lateral cutting on the west side of this bourne is marked by a slight but distinct feature, which nearly coincides with the line of the road running northward from Mill End.

Only three good sections of the chalky gravel forming these fans have been observed. One, on the west side of Oaken Grove, west of Greenlands, shows about 15 ft. of obscurely-bedded coarse gravel, consisting of angular and sub-angular flints, with an admixture of ordinary quartzite gravel, in a compact matrix of small pieces of chalk and light-brown chalky-loam, which—as Mr. Whitaker has remarked[1]—looks "very like a coarse mortar." In the upper part of this section—from 2 to 4 ft. below the surface of the ground—the drift has weathered irregularly into a dark-brown stony loam. A pit at the Greenlands gasworks showed 15 ft. of a similar drift; and another working, a little to the north of Mill End, and on the west side of the Hambleden road, shows about 12 ft. of the stony rubble, which there contains a smaller proportion of quartzites and other materials from the deposits of Plateau Gravel on the adjacent high ground.

A little stratified sandy gravel, with pebbles of chalk, was noticed, in 1899, in a pit near the ferry south-east of Mill End. This looked more like a river drift than a combe-gravel.

Chalky gravel is exposed in the banks of the lane north of Harpsden Court; and a mass of mingled fine and coarse chalk-rubble with bands and lenses of unworn flints (recalling the "Elephant Bed" of Brighton) is cut through by the Wargrave road, on the right bank of the Thames, at Marsh Mill.

In Hurley Bottom, by the third milestone out of Henley on the Marlow road (south of the river), there is a pit about 8 ft. deep "in gravel made up of flints, flint-pebbles, and pieces of chalk, the last in great abundance"[2]; and a similar gravel has been seen in the grounds of Culham Court. Many remains of mammalia, including elephant, rhinoceros, hippopotamus, and reindeer, have been found in the same kind of drift near Hurley, at a point a little beyond the eastern limit of this district.[3]

The very chalky drift above noticed thins out up the sides of the main valley, and passes downward beneath the Alluvium. It is seldom seen in the river banks, save where the stream or its backwaters run close along the foot of steep chalk slopes; and the gravel forming the whale-backed "islands" which rise through the flood loam in the water-meadows—as at Medmenham and to the west of Hambleden Lock—is generally devoid of calcareous matter, in its upper parts, at least. This chalk gravel, however, is known to follow the sides of the valley below river-level,[4] and is probably continuous with the similar drift commonly

1 'The Geology of London' (*Mem. Geol. Surv.*), 1889, vol. i, p. 388. Mr. Whitaker saw "15 to 18 ft." of drift in the Oaken Grove pit.

2 Whitaker, *loc. cit.*

3 Whitaker, *op. cit.*, p. 389.

4 Recent drainage excavations at Wargrave, 2¾ miles S.S.E. of Henley, have shown that the older part of that village, near the river front, stands on a thick spread of slope-wash, consisting of coarse sandy loam with small pellets of chalk above, and flint and chalk rubble with much loamy matter below. The drift was not bottomed in the main street, but is known to be 25 ft. thick in a well at the "Thatched House," at the N. end of the village. (*See* 'Water Supply of Berkshire,' p. 90.)

found at the base of the low-level deposits beneath and near the bed of the stream. The rubbly "bottom gravel" just mentioned seems to have been encountered in a well-boring at Messrs. Brakspear's Brewery at Henley, a little to the north of the Parish Church. The boring passed through—

| | | Ft. |
|---|---|---|
| Valley Gravel. | Made ground and gravel | 22 |
| | Sand and gravel | 4½ |
| | Gravel and chalk | 4½ |
| | Chalk | 129 [1] |

It is probable that the pebbles of chalk in the loose ballast dredged from the Thames in this neighbourhood are much more often derived from this rubbly bed than from the solid rock directly.

The talus and fan-shaped deposits of rubble, so well developed on the sides of the Thames Valley near Henley, give evidence of a rock-wasting which differed from that now in progress in being to a much greater extent a mechanical process, and withal a more rapid one. At the present day, rain falling on the chalk slopes is entangled in the vegetation, absorbed, and evaporated before it can gather into runnels of sufficient volume to effect an appreciable amount of erosion, even of the soil over the decaying rock; but when the rubble was formed, water—whether of rain or of melting snow—must, at times, have flowed freely down the steep places, furrowing the Chalk and concentrating in heavily-loaded torrents in the combes.

These deposits, which, under the existing climatic conditions, are slowly shrinking and changing into stony loams through the dissolution of their calcareous constituents, were, doubtless, formed in the same manner, and at about the same (late Pleistocene) epoch, as the banks of "Head," and wider spreads of "Coombe-rock," on the south coast.[2]

The loamy flint-gravels in the Hambleden Valley, and in the upper branches of the Wye Valley (east and south-east of Radnage), are not well exposed. Small sections in the similar drift of the Assendon Valley are to be seen at the Sewage Works, west of Round Hill.

A well at the Henley Waterworks, in the little combe to the south-west of the town, traversed—

| | | Ft. |
|---|---|---|
| Soil and Valley Gravel. | Earth and mould | 2 |
| | Clay and stones | 6 |
| | Gravel | 7 |
| | Chalk | 225 |

The well-head is about 22 ft. above the Thames. Mr. Blake was informed that the water-level is, approximately, 6½ ft. below the surface. There is an intermittent spring close by.

In the Harpsden Valley, flint-gravel, with some admixture of material from the adjacent deposits of Plateau Gravel, has been worked to a depth of 10 ft., or more, to the west of Rotherfield Peppard. Three quarters of a mile west of Harpsden Church there is a low terrace of gravel and loam

1 From Mr. J. H. Blake's MS. notes. The water-level is stated to be 9½ ft. down—*i.e.* about river-level.

2 *Vide* C. Reid, 'On the Origin of Dry Chalk Valleys and of Coombe Rock.' *Quart. Journ. Geol. Soc.*, vol. xliii, 1887, pp. 364-371.

on the northern side of the combe (Fig. 12—see next page). A pit in it showed (in 1895) the following section :—

| | | Ft. |
|---|---|---|
| Brown loam - - - - - - - - - - - | to | 20 |
| Loam, with seams of very fine gravel - - - - - - - | | $1\frac{1}{2}$—$\frac{1}{2}$ |
| Brown sandy loam - - - - - - - - - - | to | 12 |
| Gravel, consisting mainly of white sub-angular and angular flints, with some flint-pebbles, quartzites, etc. Small bodies of chalk-rubble occurred in places, and in some of them the stones were concreted into irregular blocks, up to 2 ft. in diameter, by a tough calcareous cement of a light brown or buff tint - - - - - - - | exposed | 3 to 10 |

From the arrangement of the beds in different parts of the excavation it was evident that the loams filled an old east-and-west channel in the gravel. The section is now obscured.

Parts of the tufaceous substance cementing the rubble in the gravel has a hardness little inferior to that of ordinary Chalk Rock. Treated with acetic acid, a sample of the hardened rubble assumed a cavernous character, the fragments of chalk being more soluble than their cement. The latter was easily decomposed by dilute hydrochloric acid, leaving sand, fine mud, chips of flint, and small pebbles of various kinds. The solution contained no appreciable amount of silica.

A scree of angular pieces of chalk and flint-nodules, firmly cemented by compact tufa into tabular masses over 2 ft. thick and of unknown but evidently considerable superficies, is exposed in section on the north side of the Chinnor Hill Chalk-Rock pit, which is described on an earlier page.

## Alluvium.

The recent fluviatile deposits of the water-meadows and marshes bordering the streams in the wider valleys of this area consist of indistinctly-laminated calcareous silts, loams, and clays, with subordinate beds of sand and gravel. Vegetable remains abound in the more argillaceous deposits lying below normal water-level, and occasionally form beds of peaty character. Shells of fresh-water and land mollusca are common in all kinds of sediment, and seams of shell-loam and shell-sand are often to be seen in the banks of streams and dikes.

The upper layers of the Alluvium are, as a rule, of finer texture than the lower, and the modern flood-loams of the meadows are commonly underlain, conformably or otherwise, by beds of sand and gravel. It is often difficult to decide whether these lower, coarse-grained sediments belong to the Holocene Alluvium, or to the Pleistocene "bottom gravels." Probably most of the gravel and gravelly sand actually underlying the modern flood-loams in the Thames Valley has been worked over by the river in post-Pleistocene times. The low, stony "islands" which rise above the marshes—sometimes to heights unattained by the highest recorded floods—are evidently shoals formed when the neighbouring streams possessed far greater volume and power than at the present day, and probably date from the close of the Pleistocene period.

In the western part of the district, the Thames alluvium occupies a narrow strip of country varying from rather less than one furlong to three furlongs in width, including the stream.

On the left bank of the river, about 800 yds. west of Long

*Fig.* 12.—*Section across the Valley west of Harpsden.*

Distance about 500 yards. Height exaggerated.

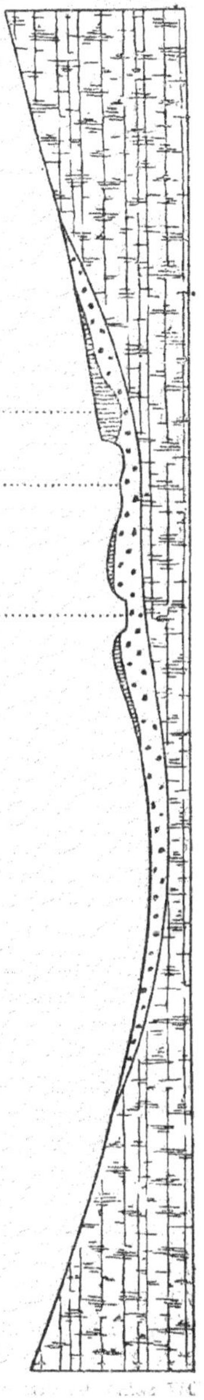


Chalk covered by Valley Gravel and Loam.

Wittenham Church, Messrs. T. Pears and L. Richardson, observed (in 1899) the following descending succession:—

"(1) Normal dark brown alluvium;
(2) Shell-marl of a greenish colour;
(3) [a] Brown argillaceous deposit literally crowded with shells;
(4) A thin accumulation of gravel containing lenticular patches of dark blue clay; and
(5) Dark blue clay (Gault or Kimmeridge Clay)."

These deposits rested in a hollow in the low-level gravel. Their thicknesses are not stated, but the height of the section is given as about 9 ft.

In samples of the earth from beds (1), (2), and (3) Messrs. A. S. Kennard and B. B. Woodward identified 54 species of freshwater and land mollusca, including:

Helix aspersa *Müll.*
Planorbis strœmii *Westld.*
Dreissensia polymorpha (*Pall.*)
Pisidium supinum *A. Schm.*

*Planorbis strœmii* and *Pisidium supinum* are believed to be extinct in the British Isles, though both were extremely common in quite recent times.

Prior to its discovery here, *Dreissensia polymorpha* was generally considered to have been introduced from the Continent within the last hundred years.

Messrs. Pears and Richardson observed a layer of shell-marl between the ordinary alluvium and the gravel, a little to the south-east of the above spot, and again at the entrance to Clifton Hampden Lock. About 300 yards to the north of Appleford Church (shown at the western edge of Sheet 254), gravelly sand crowded with shells and capped by alluvium was seen. At the base of the alluvium (*i.e.* the flood-loam) remains of dog, horse, pig, ox (*Bos longifrons* Owen), sheep, and goose were obtained, and with them was much charred wood.

Low-level Valley Gravel of fine texture and markedly current-bedded is exposed in shallow workings near the "Barley Mow" Inn, on the right bank of the Thames, to the south of Clifton Hampden Church. It consists very largely of local Greensand debris and rolled Jurassic fossils "such as *Montlivaltia*, *Thecosmilia*, *Gryphœa*, *Belemnites*, and *Rhynchonella*."[2]

The Rev. J. C. Clutterbuck found the following species of molluscs in thin laminated sandy loam, "5 ft. under gravel," near Wittenham:—

"Bithynia tentaculata, in plenty.
Pupa muscorum, in plenty.
Helix rufescens.
Valvata cristata?
Pisidium amnicum."[3]

Prof. J. Phillips has figured and described two palæoliths and three neoliths which were dredged with gravel from the bed of the river at the same locality.[4]

On the left bank of the Thames, about 100 yards north of Wallingford

1 'Holocene Deposits at Clifton Hampden, near Oxford,' Part 1, The Geology, by T. Pears and L. Richardson: Part 2, Zoology: (A) *Mollusca*, by A. S. Kennard and B. B. Woodward, (B) *Vertebrata*, by M. A. C. Hinton, *Proc. Cottesw. Nat. Fd. Club.* vol. xiv, pt. 3 (1903), pp. 195-204.

2 T. Pears and L. Richardson, on 'The Holocene Deposits at Clifton Hampden, near Oxford.' *Proc. Cottesw. Nat. Fd. Club*, vol. xiv (1903), p. 196.

3 J. Phillips, in 'The Geology of Oxford and the Valley of the Thames,' 1871, p. 467.

4 *Ibid.*, pp. 476-478.

Fig. 13.—*Section of Alluvium at Wallingford.*

Vertical Scale, 1 in. = 12 ft.

Ground liable to floods.

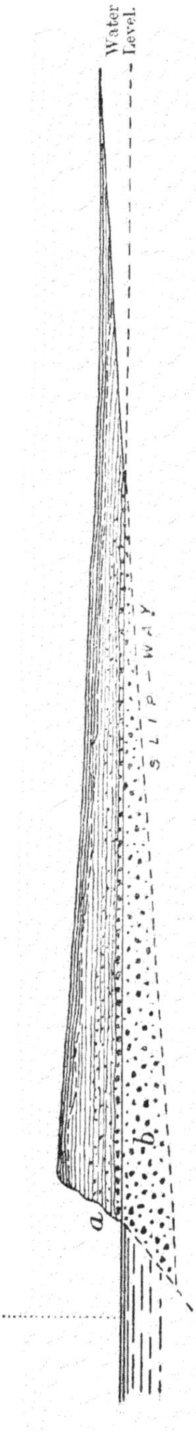


*a.* Alluvium: Dark brown sandy loam. *b.* Valley Gravel.

Bridge, an excavation for a launch-slip presented, in 1906, the following section (Fig. 13):—

| | Ft. |
|---|---|
| (*a*) Dark brown sandy loam, containing vegetable remains and penetrated by living rootlets; passing down into lighter brown, stratified, shelly loam, silty and gravelly in the lower part - to | 4½ |
| (*b*) Gravel, of very fine to medium texture and of greyish colour; consisting of small flat pieces of malmstone, sub-angular flints, quartz and quartzite pebbles, and rolled Jurassic fossils; and containing many Recent shells, with remains of horse, sheep, and ox? . . . visible for a few inches above, and a few feet below, water-level. | |

The natural embankment formed by the thickening of the loams towards the stream was well shown in the cutting. The visible portions of the gravel here evidently form part of the Alluvium, and Mr. Kennard, who has kindly examined the molluscan and other remains obtained, states that it is certainly of Holocene age.

*Planorbis strœmii* Westld. is a common form in the loams (*a*); and *Bithynia tentaculata* (Linn.), *Limnœa pereger* (Müll.), *Neritina fluviatilis* (Linn.), *Valvata piscinalis* (Müll.) are plentiful in both gravel and loams.

Part of the low-lying Valley Gravel mapped on the right bank of the river, opposite Henley, and to the north of the bridge, also belongs to the Pleistocene deposits, a deep trench, round a plantation on the west side of Remenham Lane, showing it passing laterally into the brown flood-loam.

Drains cut in the meadows to the south of Greenlands and to the north of Aston expose from 2 to 5 ft. of loam and clay, resting evenly on decalcified gravel. A band of light, shelly sand occurs in the low river-bank, on the right-hand side of the stream, a little to the north-west of Hambleden Lock; and a similar deposit is in course of formation on the shelving bottom close by.

A good section of distinctly-stratified loam with fine gravelly seams was exposed in the high bank on the right side of the river, just below Hambleden Lock, in 1904, when the protective camp-sheathing was being renewed.

Near Medmenham, the modern flood-plain of the river is nearly three-quarters of a mile wide. Much of the loam exposed in the sides of drains on the northern side of the stream here is dark and peaty.

Near Frogmill Farm, about three furlongs south-east of Medmenham Abbey, the flood-loam thins out against a mass of (Pleistocene?) gravel, which forms the high bank on the south side of the river.

The broadest spread of Alluvium in the district occurs in the marshes drained by the Mill Brook and by a smaller stream joining the Thames near Cholsey Mill, to the west and south-west of Wallingford. A good deal of the ground here is peaty.

No features of interest have been observed in the alluvium of the Thame and the Cuxham Brook. Sections of the Recent deposits of the former stream are to be seen at Overy, near Dorchester, and by the bridge at Chislehampton.

## APPENDIX.

### List of Principal Works on the Geology of the District.


1821. Buckland, Rev. Prof. W.—Description of the Quartz Rock of the Lickey Hill . . . with considerations on the evidences of a Recent Deluge afforded by the gravel beds of Warwickshire and Oxfordshire, and the Valley of the Thames from Oxford downwards to London. *Trans. Geol. Soc.*, vol. v., p. 506.

1823 ——————Reliquiæ Diluvianæ; or Observations on the organic remains contained in Caves, Fissures, and Diluvial Gravel, and on other geological phenomena, attesting the action of an universal Deluge. 4to. London. 2nd Edition. 1824.

1829. Conybeare, Rev. W. D.—On the Hydrographical Basin of the Thames, with a view more especially to investigate the causes which have operated in the formation of the valleys of that river, and its tributary streams. *Proc. Geol. Soc.*, i., 145—149.

1836. Fitton, Dr. W. H.—Observations on some of the Strata between the Chalk and the Oxford Oolite, in the South-East of England. *Trans. Geol. Soc.*, ser. 2, vol. iv., pp. 103—388.

1854. Prestwich, [Sir] J.—The Woolwich and Reading Series. *Quart. Journ. Geol. Soc.*, vol. x., p. 75.

1860. Phillips, Prof. J.—Notice of some Sections of the Strata near Oxford. *Quart. Journ. Geol. Soc.*, vol. xvi., pp. 307—311.

Geological Survey Map, Sheet 13. By W. T. Aveline, E. Hull, W. Whitaker, and others.

1861. Clutterbuck, Rev. J. C.—On the Course of the Thames from Lechlade to Windsor, as ruled by the Geological Formations over which it passes. *Rep. Brit. Assoc.* for 1860, Sections, p. 75.

Hull, E., and W. Whitaker.—The Geology of Parts of Oxfordshire and Berkshire. *Mem. Geol. Surv.* 8vo. London.

Whitaker, W.—On the "Chalk-rock," the Topmost Bed of the Lower Chalk in Berkshire, Oxfordshire, Buckinghamshire, &c. *Quart. Journ. Geol. Soc.*, vol. xvii., p. 166.

Geological Survey Map, Sheet 7. By W. Whitaker, T. R. Polwhele, and others.

Drift Edition. By W. Whitaker, H. B. Woodward, F. J. Bennett, J. H. Blake, and others. 1872.

1864. Whitaker, W.—The Geology of parts of Middlesex, Hertfordshire, Buckinghamshire, Berkshire, and Surrey. *Mem. Geol. Surv.*

1865. —————— On the Chalk of Buckinghamshire. *Quart. Journ. Geol. Soc.*, vol. xxi., p. 398.

1867. Hull, [Prof.] E., and W. Whitaker.—Horizontal Section, Sheet 71, from Nettlebed Hill, Oxfordshire. . . . *Geol. Surv.*

1871. Phillips, Prof. J.—Geology of Oxford and the Valley of the Thames, 8vo. Oxford.

1872. Claypole, E. W.—On some Gravels in the Valley of the Thames in Berkshire. *Proc. Bristol Nat. Soc.*, ser. 2, vol. vi., pp. 28—33.

Whitaker, W.—The Geology of the London Basin. *Mem. Geol. Surv.*, vol. iv.

1876. Barrois, [Prof.] C.—Recherches sur le terrain Crétacé supérieur de l'Angleterre et de l'Irlande. *Mem. Soc. Géol. du Nord*, 4to. Lille.

1883. WHITAKER, W.—List of Works on the Geology and Palæontology of Oxfordshire, of Berkshire, and of Buckinghamshire. *Rep. Brit. Assoc.* for 1882, p. 327.

1885. HINDE, Dr. G. J.—On Beds of Sponge Remains in the Lower and Upper Greensand of the South of England. *Phil. Trans.*, vol. clxxvi.

IRVING, Rev. A.—Water Supply from the Bagshot and other Strata. *Geol. Mag.*, pp. 22, 23.

1889. WHITAKER, W.—The Geology of London and of part of the Thames Valley. *Mem. Geol. Surv.* 2 vols.

1890. PRESTWICH, [Sir] J.—On the relation of the Westleton Beds, etc. Three parts. *Quart. Journ. Geol. Soc.*, vol. xlvi., pp. 84, 120, 155.

1891. MONCKTON, H. W., and R. S. HERRIES.—On some Hill Gravels North of the Thames. *Proc. Geol. Assoc.*, vol. xii., pp. 108—114.

1892. WHITE, H. J. O.—Notes on the Westleton Beds near Henley-on-Thames. *Proc. Geol. Assoc.*, vol. xii., pp. 379—384.

1895. WHITE, H. J. O.—On the Distribution and Relations of the Westleton and Glacial Gravels in parts of Oxfordshire and Berkshire. *Proc. Geol. Assoc*, vol. xiv., pp. 11—30.

WOODWARD, H. B.—Jurassic Rocks of Britain, vol. v. Middle and Upper Oolitic Rocks of England (Yorkshire excepted). [With Bibliography relating to Jurassic formations.] *Mem. Geol. Surv.* 8vo. London.

1896. SALTER, Dr. A. E.—Pebbly Gravel—Goring Gap to Norfolk Coast. *Proc. Geol. Assoc.*, vol. xiv., pp. 389—404.

1897. WHITE, H. J. O.—Origin of High-level Gravel with Triassic Débris adjoining the Valley of the Upper Thames. *Proc. Geol. Assoc.*, vol. xv., pp. 157—174.

1898. SALTER, A. E.—Pebbly and other Gravels in Southern England. *Proc. Geol. Assoc.*, vol. xv., p. 264.

SHRUBSOLE, O. A.—On some High-level Gravels in Berkshire and Oxfordshire. *Quart. Journ. Geol. Soc.*, vol. liv., pp. 585—600.

1899. DAVIES, A. MORLEY.—Contributions to the Geology of the Thame Valley. *Proc. Geol. Assoc.*, vol. xvi., pp. 15—58.

WHITE, H. J. OSBORNE.—Cycling Excursion to the Chiltern Hills. *Proc. Geol. Assoc.*, vol. xvi., pp. 251—254.

1900-4. JUKES-BROWNE, A. J.—The Cretaceous Rocks of Britain, vol. i. The Gault and Greensand, vol. ii. The Lower and Middle Chalk, 1903, vol. iii. The Upper Chalk, 1904. *Mem. Geol. Surv.*

1902. BLAKE, J. H., and W. WHITAKER.—The Water Supply of Berkshire from Underground Sources. *Mem. Geol. Surv.*

WHITE, H. J. O.—On a Peculiarity in the Course of Certain Streams in the London and Hampshire Basins. *Proc. Geol. Assoc.*, vol. xvii., p. 399—413.

1903. PEARS, T., and L. RICHARDSON on Geology; and A. S. KENNARD, B. B. WOODWARD, and M. A. C. HINTON on Zoology of Holocene Deposits at Clifton Hampden, near Oxford. *Proc. Cotteswold Nat. Club*, vol. xiv., pp. 195—204.

WHITE, H. J. O.—Excursion to Culham and Wallingford, *Proc. Geol. Assoc.*, vol. xviii., pp. 300—306.

1904. GREEN, Prof. A. H.—Some Points of Special Interest in the Geology of the Neighbourhood of Oxford. *Rep. Brit. Assoc.* for 1904, pp 644—645.

WHITE, H. J. OSBORNE.—Excursion to Henley-on-Thames. *Proc. Geol. Assoc.*, vol. xviii., pp. 414—418.

1905. Lamplugh, G. W., and T. I. Pocock.—Notes on the Oxford Area. *Summary of Progress of Geol. Surv.* for 1904, pp. 17—22.

Monckton, H. W., and O. A. Shrubsole.—Excursion to Reading and Caversham. *Proc. Geol. Assoc.*, vol. xix., pp. 135—137.

Salter, Dr. A. E.—On the Superficial Deposits of Central and parts of Southern England. *Proc. Geol. Assoc.*, vol. xix., pp. 1—56.

1906. Dawkins, C. G. E.—Discovery of *Exogyra sinuata* in the Lower Greensand of Culham, near Oxford. *Geol. Mag.*, p. 94.

Jukes-Browne, A. J.—The Clay-with-Flints; its Origin and Distribution. *Quart. Journ. Geol. Soc.*, vol. lxii., p. 132.

Kennard, A. S., and B. B. Woodward.—On Sections in the Holocene Alluvium of the Thames at Staines and Wargrave. *Proc. Geol. Assoc.*, vol. xix., pp. 252—258.

White, H. J. O.—On the Occurrence of Quartzose Gravel in the Reading Beds at Lane End, Bucks. *Proc. Geol. Assoc.*, vol. xix., pp. 371—377.

1907. Harmer, F. W.—On the Origin of Certain Cañon-like Valleys associated with Lake-like Areas of Depression. *Quart. Journ. Geol. Soc.*, vol. lxiii., pp. 470—513.

White, H. J. O.—Excursion to Lane End, Bucks. *Proc. Geol. Assoc.*, vol. xx., pp. 101—103.

1908. Pocock, T. I., with contributions by H. B. Woodward and G. W. Lamplugh.—The Geology of the Country around Oxford. *Mem. Geol. Surv.*

Chatwin, C. P., and T. H. Withers.—The Zones of Chalk in the Thames Valley between Goring and Shiplake. *Proc. Geol. Assoc.*, vol. xx., p. 390.

# INDEX.

## GENERAL MEMOIRS.

SUMMARY OF PROGRESS of the GEOLOGICAL SURVEY for 1897 to 1907. Each 1*s*.

PLIOCENE DEPOSITS of BRITAIN. By C. REID. 5*s*. 6*d*.

CRETACEOUS ROCKS OF BRITAIN.—VOL. I. GAULT AND UPPER GREENSAND OF ENGLAND. 9*s*. Vol. II. LOWER AND MIDDLE CHALK. 10*s*. VOL. III. UPPER CHALK. 10*s*. By A. J. JUKES-BROWNE and W. HILL.

JURASSIC ROCKS OF BRITAIN.—Vol. I. YORKSHIRE, 8*s*. 6*d*. Vol. II. YORKSHIRE, Fossils, 12*s*. By C. FOX-STRANGWAYS. Vol. III. LIAS OF ENGLAND (Yorkshire excepted). 7*s*. 6*d*. By H. B. WOODWARD. Vol. IV. The Lower Oolitic Rocks of England. 10*s*. By H. B. WOODWARD. Vol. V. The Middle and Upper Oolitic Rocks of England. 7*s*. 6*d*. By H. B. WOODWARD.

BRITISH ORGANIC REMAINS. DECADES I. to XIII., with 10 Plates each. Price 4*s*. 6*d*. each 4to; 2*s*.6*d*. each 8vo.

MONOGRAPH I. On the Genus PTERYGOTUS. By T. H. HUXLEY and J. W. SALTER. 7*s*.

MONOGRAPH II. On the Structure of the BELEMNITIDÆ. By T. H. HUXLEY. 2*s*. 6*d*.

MONOGRAPH III. CROCODILIAN REMAINS found in the ELGIN SANDSTONES. By T. H. HUXLEY. 14*s*. 6*d*.

MONOGRAPH IV. On the CHIMÆROID FISHES of the British Cretaceous Rocks. By E. T. NEWTON. 5*s*.

VERTEBRATA of the PLIOCENE DEPOSITS of BRITAIN. By E. T. NEWTON. 4*s*.

*Museum Catalogues, &c.:—*

HANDBOOK to BRITISH MINERALS. By F. W. RUDLER. 1*s*.

HANDBOOK to the MUSEUM of PRACTICAL GEOLOGY. 6*d*.

FOSSILS:—CAMBRIAN and SILURIAN, 2*s*. 6*d*.; CRETACEOUS, 2*s*. 9*d*.; TERTIARY and POST-TERTIARY, 1*s*. 8*d*.

## DISTRICT MEMOIRS.

CAMBRIDGE. By W. H. PENNING and A. J. JUKES-BROWNE. 4*s*. 6*d*.

CORNWALL, DEVON, AND WEST SOMERSET. INDEX to DE LA BECHE'S Report on. By C. REID. 1*s*.

DERBYSHIRE, NORTH. By A. H. GREEN, C. LE NEVE FOSTER and J. R. DAKYNS. 2nd Ed. By A. H. GREEN and A. STRAHAN. 5*s*. 6*d*.

FALMOUTH AND TRURO, and MINING DISTRICT OF CAMBORNE AND REDRUTH. By J. B. HILL and D. A. MACALISTER. 7*s*. 6d.

FENLAND. By S. B. J. SKERTCHLY. 36*s*. 6*d*.

HOLDERNESS. By C. REID. 4*s*.

ISLE OF MAN. By G. W. LAMPLUGH. 12*s*.

TERTIARY FLUVIO-MARINE FORMATION of the ISLE OF WIGHT. By EDWARD FORBES. 5*s*.

ISLE OF WIGHT. By H. W. BRISTOW. New Ed. By C. REID and A. STRAHAN. 8*s*. 6*d*.

ISLE OF PURBECK AND WEYMOUTH. By A. STRAHAN. 10*s*. 6*d*.

GUIDE TO GEOLOGICAL MODEL OF ISLE OF PURBECK. By A. STRAHAN. 6*d*.

LAKE DISTRICT, NORTHERN PART OF. By J. C. WARD. 9*s*.

LANCASHIRE, SUPERFICIAL DEPOSITS OF SOUTH WEST. By C. E. DE RANCE. 10*s*. 6*d*.

LONDON AND PART OF THE THAMES VALLEY. By W. WHITAKER. Vol. I., 6*s*. Vol. II. 5*s*.

MIDLAND COUNTIES, TRIASSIC and PERMIAN ROCKS of the. By E. HULL. 5*s*.

NORFOLK and SUFFOLK, VERTEBRATA of the FOREST BED SERIES of. By E. T. NEWTON. 7*s*. 6*d*.

NORWICH. By H. B. WOODWARD. 7*s*.

RUTLAND, &c. By J. W. JUDD. 12*s*. 6*d*.

*Eruptive Rocks:—*

ERUPTIVE ROCKS of BRENT TOR. By F. RUTLEY. 15*s*. 6*d*.

FELSITIC LAVAS of ENGLAND and WALES. By F. RUTLEY. 9*d*

*Iron Ores of Great Britain:—*

Part II. South Staffordshire. Price 1*s*. Part III. South Wales. Price 1*s*. 3*d*. Part IV. The Shropshire Coalfields and North Staffordshire. 1*s*. 3*d*.

*Coal Fields:—*

YORKSHIRE COALFIELD. By A. H. GREEN, R. RUSSELL [and Others]. 42*s*.

EAST SOMERSET and BRISTOL COALFIELDS. By H. B. WOODWARD. 18*s*.

WARWICKSHIRE COALFIELD. By H. H. HOWELL. 1*s*. 6*d*.

LEICESTERSHIRE and SOUTH DERBYSHIRE COALFIELD. By C. FOX-STRANGWAYS. 6*s*.

NORTH STAFFORDSHIRE COALFIELDS. By W. GIBSON [and Others]. 6*s*.

SOUTH WALES, THE COALS OF. By A. STRAHAN and W. POLLARD. 1*s*. 6*d*.

SOUTH WALES. (See also Sheet Memoirs, New Series).

*Sanitation and Water Supply:—*

SOILS AND SUBSOILS from a SANITARY POINT of VIEW. By H. B. WOODWARD. 2nd Ed. 1*s*. 6*d*.

WATER SUPPLY OF SUSSEX. By W. WHITAKER and C. REID. 3*s*.

WATER SUPPLY OF BERKSHIRE. By J. H. BLAKE and W. WHITAKER. 3*s*.

WATER SUPPLY OF LINCOLNSHIRE. By H. B. WOODWARD [and Others] 4*s*. 6*d*.

WATER SUPPLY OF SUFFOLK. By W. WHITAKER [and Others]. 3*s*. 6*d*.

WATER SUPPLY OF EAST RIDING OF YORKSHIRE. By C. FOX-STRANGWAYS, with contributions by H. R. MILL. 3*s*.

## SHEET MEMOIRS OF OLD SERIES MAPS.

London Sheet. Guide to the GEOLOGY of LONDON and NEIGHBOURHOOD. By W. WHITAKER. 6th Ed. 1*s*.

34 - - - PARTS of WILTS, and GLOUCESTERSHIRE. By A. C. RAMSAY, W. T. AVELINE, and E. HULL. 8*d*.

44 - - - CHELTENHAM. By E. HULL. 2*s*. 6*d*.

45 - - - BANBURY, WOODSTOCK, and BUCKINGHAM. By A. H. GREEN. 2*s*.

45 SW - WOODSTOCK. By E. HULL. 1*s*.

47 - - - N.W. ESSEX & N.E. HERTS. By W. WHITAKER, W. H. PENNING, W. H. DALTON, & F. J. BENNETT. 2*s*. 6*d*.

48 SW - - COLCHESTER. By W. H. DALTON. 1*s*. 6*d*.

48 SE - - EASTERN END of ESSEX (WALTON NAZE and HARWICH). By W. WHITAKER. 9*d*.

48 NW, NE - IPSWICH, HADLEIGH, and FELIXSTOWE. By W. WHITAKER, W. H. DALTON, & F. J. BENNETT. 2*s*.

49 S, 50 SE - ALDBOROUGH, &c. By W. H. DALTON. Edited, with additions, by W. WHITAKER. 1*s*.

49 N - - SOUTHWOLD. By W. WHITAKER. 2*s*. 6*d*.

50 SW - STOWMARKET. By W. WHITAKER, F. J. BENNETT, and J. H. BLAKE. 1*s*.

50 NW - - DISS, EYE, &c. By F. J. BENNETT. 2*s*.

50 NE - - HALESWORTH and HARLESTON. By W. WHITAKER and W. H. DALTON. 1*s*.

51 SE - - BURY ST. EDMUNDS and NEWMARKET. By F. J. BENNETT, J. H. BLAKE, and W. WHITAKER. 1*s*.

51 NE - PARTS of CAMBRIDGESHIRE and SUFFOLK. By W. WHITAKER [and Others]. 2*s*.

53 SE - PART of NORTHAMPTONSHIRE. By W. T. AVELINE and RICHARD TRENCH. 8*d*.

53 NE - - PARTS of NORTHAMPTONSHIRE and WARWICKSHIRE. By W. T. AVELINE. *8d.*
65 - - - S.W. NORFOLK and N. CAMBS. By W. WHITAKER, S. B. J. SKERTCHLY and A. J. JUKES-BROWNE. *3s.*
66 SW - - ATTLEBOROUGH. By F. J. BENNETT. *1s. 6d.*
67 NW - - E. DEREHAM. By J. H. BLAKE. *1s. 6d.*
67 - - YARMOUTH and LOWESTOFT. By J. H. BLAKE. *2s.*
68 E - - CROMER. By C. REID. *6s.*
68 NW, SW - FAKENHAM, WELLS, &c. By H. B. WOODWARD. *2s.*
69 - - BORDERS OF THE WASH. By W. WHITAKER and A. J. JUKES-BROWNE. *3s.*
70 - - S. W. LINCOLNSHIRE, &c. By A. J. JUKES-BROWNE and W. H. DALTON. *4s.*
71 NE - - NOTTINGHAM. By W. T. AVELINE. (2nd Ed.) *1s.*
79 NW - - RHYL, ABERGELE, and COLWYN. By A. STRAHAN. (Notes by R. H. TIDDEMAN). *1s. 6d.*
79 SE - FLINT, MOLD, and RUTHIN. By A. STRAHAN. (Parts by C. E. DE RANCE). *4s. 6d.*; *Supplement 2d.*
80 NW - - PRESCOT, LANCASHIRE. By E. HULL. (3rd Ed. With additions by A. STRAHAN.) *3s.*
80 SW - - CHESTER. By A. STRAHAN. *2s.*
81 NW, SW - STOCKPORT, MACCLESFIELD, CONGLETON, and LEEK. By E. HULL and A. H. GREEN. *4s.*
82 SE - - PARTS of NOTTINGHAMSHIRE and DERBYSHIRE. By W. T. AVELINE. (2nd Ed.) *6d.*
82 NE - - PARTS of NOTTINGHAMSHIRE, YORKSHIRE, and DERBYSHIRE. (2nd Ed.) By W. T. AVELINE. *1s.*
83 - - LINCOLN. By W. A. E. USSHER, A. J. JUKES-BROWNE, and A. STRAHAN. *3s.*
84 - - EAST LINCOLNSHIRE. By A. J. JUKES-BROWNE. *3s. 6d.*
86 - - N. LINCOLNSHIRE and S. YORKSHIRE. By W. A. E. USSHER [and Others]. *2s.*
87 NW - - WAKEFIELD and PONTEFRACT. By A. H. GREEN. *6d.*
87 SW - - BARNSLEY. By A. H. GREEN. *9d.*
88 SW - - OLDHAM. By E. HULL. *2s.*
88 SE - - PART of the YORKSHIRE COALFIELD. By A. H. GREEN, J. R. DAKYNS, and J. C. WARD. *1s.*
88 NE - - DEWSBURY, &c. By A. H. GREEN, J. R. DAKYNS, J. C. WARD, and R. RUSSELL. *6d.*
89 SE - - BOLTON, LANCASHIRE. By E. HULL. *2s.*
90 SE - - COUNTRY between LIVERPOOL and SOUTHPORT. By C. E. DE RANCE. *3d.*
90 NE - - SOUTHPORT, LYTHAM, and SOUTH SHORE. By C. E. DE RANCE. *6d.*
91 SW - - COUNTRY between BLACKPOOL and FLEETWOOD. By C. E. DE RANCE. *6d.*
91 NW - - SOUTHERN PART of the FURNESS DISTRICT in N. LANCASHIRE. By W. T. AVELINE. *6d.*
92 SE - - BRADFORD and SKIPTON. By J. R. DAKYNS, C. FOX-STRANGWAYS, R. RUSSELL, and W. H. DALTON. *6d.*
93 NW - - NORTH and EAST of HARROGATE. By C. FOX-STRANGWAYS. *6d.*
93 NE - - COUNTRY between YORK and MALTON. By C. FOX-STRANGWAYS. *1s. 6d.*
93 NW - - N. and E. of LEEDS and near TADCASTER. By W. T. AVELINE, A. H. GREEN [and Others].
93 SE, 94 SW COUNTRY between YORK and HULL. By J. R. DAKYNS, C. FOX-STRANGWAYS, and A. G. CAMERON. *1s. 6d.*
94 NW - - DRIFFIELD. By J. R. DAKYNS and C. FOX-STRANGWAYS. *9d.*
94 NE - - BRIDLINGTON BAY. By J. R. DAKYNS and C. FOX-STRANGWAYS. *1s.*
95 SW, SE - SCARBOROUGH and FLAMBOROUGH HEAD. By C. FOX-STRANGWAYS. (2nd Ed.) *4s. 6d.*
95 NW WHITBY and SCARBOROUGH. By C. FOX-STRANGWAYS and G. BARROW. *1s. 6d.*
96 SE - - NEW MALTON, PICKERING, and HELMSLEY. By C. FOX-STRANGWAYS. *1s.*
96 NE - - ESKDALE, ROSEDALE, &c. By C. FOX-STRANGWAYS, C. REID, and G. BARROW. *1s. 6d.*
96 NW, SW - NORTHALLERTON and THIRSK. By C. FOX-STRANGWAYS, A. G. CAMERON, and G. BARROW *1s. 6d.*
97 SW - - INGLEBOROUGH. By J. R. DAKYNS, R. H. TIDDEMAN, W. GUNN, and A. STRAHAN. *2s.*
97 NW - - MALLERSTANG. By J. R. DAKYNS, R. H. TIDDEMAN [and Others]. *3s. 6d.*
98 NE - - KENDAL. By W. T. AVELINE and T. MC K. HUGHES. 2nd Ed. by A. STRAHAN. *2s.*
102 SW - - APPLEBY, ULLSWATER, &c. By J. R. DAKYNS, R. H. TIDDEMAN, and J. G. GOODCHILD. *1s. 6d.*
104 SW, SE - NORTH CLEVELAND. By G. BARROW. *1s. 6d.*
107 - - CARLISLE. By T. V. HOLMES. *1s. 3d.*
108 SE - - OTTERBURN and ELSDON. By HUGH MILLER. *2s. 6d.*
108 NE - - CHEVIOT HILLS. By C. T. CLOUGH. *1s. 6d.*
108 SW - - PLASHETTS and KIELDER. By C. T. CLOUGH. *1s.*
110 SW - WOOLER and COLDSTREAM. By W. GUNN and C. T. CLOUGH. *1s. 6d.*
110 NW - - NORHAM and TWEEDMOUTH. By W. GUNN. *6d.*
110 NE - - COAST SOUTH of BERWICK-ON-TWEED. By W. GUNN. *9d.*
110 SE - - BELFORD, HOLY ID, and FARNE ISLANDS. By W. GUNN. *2s. 6d.*

## SHEET MEMOIRS OF NEW SERIES MAPS.

110 - - MACCLESFIELD, CONGLETON, &c. By T. I. POCOCK [and Others]. *2s. 6d.*
123 - - STOKE-UPON-TRENT. By W. GIBSON and C. B. WEDD. (2nd Ed.) Price *1s. 6d.*
125 - - - DERBYSHIRE and NOTTINGHAMSHIRE COALFIELD (Southern part). By W. GIBSON, &c. *3s.*
141 - - - DERBY, BURTON-ON-TRENT, &c. By C. FOX-STRANGWAYS. *2s.*
155 - - - ATHERSTONE and CHARNWOOD FOREST. By C. FOX-STRANGWAYS. *2s.*
156 - - - LEICESTER. By C. FOX-STRANGWAYS. *3s.*
230 - - - AMMANFORD. By A. STRAHAN [and Others]. *2s. 6d.*
231 - - - MERTHYR TYDFIL. By A. STRAHAN, W. GIBSON, and T. C. CANTRILL. *1s. 6d.*
232 - - - ABERGAVENNY. By A. STRAHAN and W. GIBSON. *2s.*
246 - - - WEST GOWER. By A. STRAHAN. *8d.*
247 - - - SWANSEA. By A. STRAHAN [and Others]. *2s. 6d.*
248 - - - PONTYPRIDD. By A. STRAHAN, R. H. TIDDEMAN, and W. GIBSON. *1s. 6d.*
249 - - - NEWPORT, MON. By A. STRAHAN. *2s.*
262 - - - BRIDGEND. By A. STRAHAN and T. C. CANTRILL. *1s. 6d.*
263 - - - CARDIFF. By A. STRAHAN and T. C. CANTRILL. *2s. 3d.*
267 - - - HUNGERFORD and NEWBURY. By H. T. OSBORNE WHITE. *2s. 6d.*
268 - - - READING. By J. H. BLAKE and H. W. MONCKTON. *1s. 6d.*
282 - - - DEVIZES. By A. J. JUKES-BROWN. *1s.*
283 - - - ANDOVER. By A. J. JUKES-BROWNE.
295 - - - QUANTOCK HILLS and TAUNTON. By W. A. E. USSHER.
298 - - - SALISBURY. By C. REID [and Others]. *1s. 3d.*
311 - - - WELLINGTON and CHARD. By W. A. E. USSHER. *1s. 3d.*
314 - - - RINGWOOD. By C. REID [and Others]. *1s.*
315 - - - SOUTHAMPTON. By C. REID. *1s. 6d.*
317 - - - CHICHESTER. By C. REID [and Others]. *1s.*
325 - - - EXETER. By W. A. E. USSHER. *2s.*
326 & 340 - SIDMOUTH and LYME REGIS. By H. B. WOODWARD and W. A. E. USSHER. *1s.*
328 - - - DORCHESTER. By C. REID. *1s.*
329 - - - BOURNEMOUTH. By C. REID. *4d.*
332 - - - BOGNOR. By C. REID. *6d.*
334 - - - EASTBOURNE. By C. REID. *6d.*
346 - - - NEWQUAY. By C. REID and J. B. SCRIVENOR. *3s.*
348 - - - PLYMOUTH and LISKEARD. By W. A. E. USSHER. *3s.*
350 - - - TORQUAY. By W. A. E. USSHER. *2s.*
351 & 358 - LAND'S END DISTRICT. By C. REID and J. S. FLETT. *3s. 6d.*
352 - - - FALMOUTH AND TRURO, and MINING DISTRICT OF CAMBORNE AND REDRUTH. By J. B. HILL and D. A. MACALISTER. *7s. 6d.*
353 - - MEVAGISSEY. By C. REID. *2s.*
355 & 356 - KINGSBRIDGE and SALCOMBE. By W. A. E. USSHER. *1s. 6d.*
357 & 360 - ISLES OF SCILLY. By G. BARROW. *1s.*
—— - - OXFORD. By T. I. POCOCK [and Others]. *2s. 3d.*

Spotted something that isn't quite right?

Let us know and we'll fix it for future reprints. And as a thank you for your diligence, we'll send you a free Sothis book of your choice.

publisher@sothis.press

This page intentionally left blank

Except of course for that line above. And this one.
And whilst we have your attention, perhaps you'd
like to follow us on Instagram @sothis.press. Go on.

Made in the USA
Monee, IL
05 November 2026

67808957R00080